FARM MANAGEMENT

FARM MANAGEMENT

Kevin Longbow

Kruger Brentt
Publishers

2023

Kruger Brentt Publishers UK. LTD.
Company Number 9728962

Regd. Office: 68 St Margarets Road, Edgware, Middlesex HA8 9UU

© 2023 AUTHOR
ISBN: 9781787150720

For information on all our publications visit our website at http://krugerbrentt.com/

PREFACE

A farm is the area cultivated by a farmer or by a group of farmers in common. Legally, a farm means an area of land under single ownership and devoted to agriculture either to raise crops or for pasture. It may be a one field or many fields. On the contrary, management refers to the function of implementing the production plan and making necessary adjustments in the business of farming so that farm business may yield maximum returns. Agriculture entails growing of crops and raising of livestock for milk, wool, eggs, meat, provision of raw materials for industries and marketing of agricultural products for man's use. However, with the passage of time farm practices have been improved to meet the growing demand and improving quality of the agricultural products. The natural inputs are being increasingly replaced by purchased inputs. Similarly, the production instead of being used for family is market oriented. Thus, the concept of agriculture is charging from just a way of life to a business preposition. There are many branches of agriculture which among others include: agricultural economics, crop science, soil science, animal science, fishery, forestry and agricultural engineering.

Farm management is a science which deals with the proper combination and operation of production factors including land, labour and capital and the choice of crop and livestock enterprises to bring about a maximum and continuous return to the most elementary operation units of farming. The term farm Management consists of two words, 'Farm' and 'Management'. 'Farm' is a modified piece of land held or operated as a unit for the production of agricultural products. 'Management' means the art of managing the farm. It involves the process of determining how the farm unit shall be organised and operated. Farm is a socio-economic unit which not only provides income to a farmer but also a source of happiness to him and his family. It is also a decision-making unit where the farmer has many alternatives for his resources in the production of crops and livestock enterprises and their disposal. Hence, the farms are the micro units of vital importance which represents centre of dynamic decision making in regard to guiding the farm resources in the production process. The welfare of a nation depends upon happenings in the organisation in each farm unit. It is clear that agricultural production of a country is the sum of the contributions of the individual farm units and the development of agriculture means the development of millions of individual farms.

The prosperity of any country depends upon the prosperity of farmers, which in turn depends upon the rational allocation of resources among various uses and adoption improved technology. Human race depends more on farm products for their existence than anything else since food, clothing – the prime necessaries are products of farming industry. Even for industrial prosperity, farming industry forms the basic infrastructure. Thus, the study farm management has got prime importance in any economy particularly on agrarian economy. Farm management is becoming popular in the present world. Its scope is very wide and increasing year after year. Agriculture is modernising and to carry out agriculture successfully, knowledge of farm management is pre-requisite. In a sense it deals with the allocation of resources at the level of an individual farm. It covers aspects of farm business which have a bearing on the efficiency of the farm. The subject of farm management consists research, teaching and extension.

The present book contains eleven chapters covering all related disciplines. These chapters include introduction, principles of farm management, farm systems, strategic management, farm management decision making process, agriculture production management, marketing, dairy farm management, crop management, organic farming management, impact of the COVID-19 pandemic on agriculture and the rural economy. It is a comprehensive basic book on farm management and designed to introduce students to the key concepts on how to effectively manage a farm business. The book is written in a very simple form with the basic information needed to measure management performance, financial progress, and the financial condition of the farm business. It provides valuable insight on the impact of the pandemic on agriculture and the rural economy. This book would be an invaluable asset for the students, teachers and researchers associated with farm management.

We are grateful to all those persons as well as various books, manuals, periodicals, magazines, journals etc. that helped in the preparation of this book. In spite of the best efforts, it is possible that some errors may have occurred into the compilation and editing of the book. Further queries, constructive suggestions and criticisms for the improvement of the book are always welcome and shall be thankfully acknowledged.

Kevin Longbow

CONTENTS

Preface		**v**
1.	**Introduction**	**1**
	1.1 Objectives	1
	1.2 Introduction	1
	1.3 Meaning of Farm Management	2
	1.4 Definitions	3
	1.5 Nature of Farm Management	3
	1.6 Importance of Farm Management	4
	1.7 Production	6
	1.8 Conservation Management	8
	1.9 Role of IT in Agriculture	12
	1.10 Significance of Awareness	13
	1.11 Farming in India	14
	1.12 Sedentary Cultivation or Permanent Agriculture	16
	1.13 Economic Principles Applied to Farm Management	18
	1.14 Relationship of Farm Management with Other Sciences	19
2.	**Principles of Farm Management**	**20**
	2 .1 Objectives	20
	2.2 Introduction	20
	2.3 Farm Management Principles	23

2.4 Determination of Optimum Production Combination by Graphic Method   29

2.5 Management Procedures in Various Farm Systems   29

3. Farm Systems   **34**

3.1 Objectives   34

3.2 Introduction   34

3.3 Definitions of Farming Systems   35

3.4 General Requirements Towards Farming System Classifications   37

3.5 General Systems Classification   37

3.6 Nature of Farm-Level Systems   43

3.7 Structural Elements of the Farm-Household System   47

3.8 Structural Model of a Farm-Household System   49

3.9 Shifting Cultivation   53

3.10 Commercial Agriculture   54

3.11 Plantation Farming   56

3.12 Crop Rotation   56

3.13 Dairy Farming   57

3.14 Alternative Bases for Farm-System Analysis   59

3.15 Analytical Situations within Modes   63

3.16 Farm Types and Structure   64

3.17 Land Use and Livestock   65

3.18 Farm Buildings   66

4. Strategic Management   **68**

4.1 Objectives   68

4.2 Introduction   68

4.3 Strategic Management for Agribusiness   69

4.4 Strategic Planning for Farm Businesses   72

4.5 Implementation and Control   78

4.6 Information Systems for Farm Management   85

5. Farm Management Decision Making Process   **87**

5.1 Objectives   87

5.2 Introduction   87

5.3 Decision Making Process   90

5.4 Decision Making Framework   93

5.5 Complexity   94

5.6 Ability to Observe   95

5.7 The Decision-Making Process   97

5.8 Choose the Best Alternative   97

5.9 Strategic Decisions for the Entire Farm   98

5.10 Adaptation for the Agricultural Season and the Farm   98

6. Agriculture Production Management   102

6.1 Objectives   102

6.2 Introduction   102

6.3 Profitability of Agriculture   105

6.4 Asymmetry of Investment and Risk   112

6.5 Factors Influencing Growth in Agriculture   115

6.6 Public Investment and Policy   117

6.7 Aggregation   118

6.8 Measuring Inputs and Outputs   119

6.9 Agriculture, Environment and Natural Resources   120

6.10 Environment and Natural Resource Management (NRM)   123

6.11 Factors of Production - The Agriculture Economic   124

6.12 Theory of Production   132

6.13 Marginal Cost and Price   134

7. Marketing   137

7.1 Objectives   137

7.2 Introduction   137

7.3 Activities and Functions   137

7.4 Strategic Models   141

7.5 Enterprise Marketing Management   143

7.6 Metrics and Management   144

7.7 Budgets as Managerial Tools   148

7.8 Marketing Channels   149

7.9 Agricultural Marketing Development 151

7.10 Market Information 152

7.11 Agricultural Value Chain 154

7.12 Co-Operatives in the Agriculture and Food Sectors 156

7.13 Agricultural and Food Marketing 160

7.14 Marketing Functions 162

7.15 Advantages and Disadvantages of Diversification in Agriculture Environmental Science 164

7.16 Crop Diversification 165

8. Dairy Farm Management **167**

8.1 Objectives 167

8.2 Introduction 167

8.3 Scope for Dairy Farming and Its National Importance 168

8.4 Digital Dairy Management 170

8.5 Vision of Digital Dairy Management: Focus on Three Key Areas 174

9. Crop Management **182**

9.1 Objectives 182

9.2 Introduction 182

9.3 Preparation of Soil 183

9.4 Post-Harvest Storage 187

9.5 Impact of Climate Change on Agriculture 191

9.6 Pest Insects and Climate Change 196

9.7 Projections of Impacts 198

9.8 Comparison of Temperate and Tropical Agriculture 200

9.9 The Influence of Climate Change on Crop Production 203

10. Organic Farming Management **208**

10.1 Objectives 208

10.2 Introduction 208

10.3 Organic Farming Techniques 209

10.4 Livestock 211

10.5 Controversy 212

10.6 Advantages and Disadvantages of Organic Farming   213

10.7 GM Crops   214

10.8 Environmental Effects of Organic Farming   215

10.9 Scope of Organic Farming   216

10.10 Organic Farming in India   217

10.11 Organic Agricultural Movements   219

10.12 Organic Farming   221

10.13 Organic Agriculture – The Experience of India   222

10.14 Organic Farming Steps to a Successful Organic Transition   225

10.15 Organic Farming India's Future Perfect   229

10.16 Food Security and Safety   234

10.17 Achieving Food Security   236

10.18 Nutrition, Food Safety and Food Security   240

10.19 India's Food Security Emergency   241

10.20 Strive for No Surprises   243

11. Impact of the Covid-19 Pandemic On Agriculture and the Rural Economy   245

11.1 Introduction: An Overview   245

11.2 Global vs National Yield of Major Crops   246

11.3 Impact of Covid-19 on Farm Gate Prices: State Level   246

11.4 Impact of Covid-19 on Availability of Agri-Inputs   247

11.5 Impact on Availability of Agri-Inputs: State Level   248

11.6 Impact of Covid-19 on Prices of Agri-Inputs: State Level   251

11.7 Impact on MSMES   254

11.8 At a Glance   256

Bibliography   **258**

Index   **262**

1

INTRODUCTION

1.1 OBJECTIVES

After studying this chapter you will be able to understand:

- Meaning Nature and Definitions of Farm Management.
- Importance of Farm Management.
- Role of Farm Management in Agriculture.
- Conservation Management.

1.2 INTRODUCTION

Farm management is of the recent origin. The term "Farm Management" conveys different meanings by different people. Some take it to be another name of production economics or agricultural economics, while others consider farm management as nothing more than the farmer's art of carrying out the daily work of supervision of farm.

Like any other economic problem, farm management as a rational resource allocation proposition more particularly from the point of view of an individual farmer. On one hand a farmer has certain set of farm resources such as land, labour, farm buildings, working capital, farm equipments etc. On the other side, the same farmer has a set of goals or objectives to achieve, may be maximum family satisfaction through increasing net farm income.

Farm management is the collective term for various management strategies and methods that are employed to keep a farm productive and profitable. The process of this type of management is often associated with large commercial farms, although many of the same methods can be used with equal success on a small family-owned farm. Depending on the size of the operation, the management process may require the services of a single farmmanager or a group of managers who oversee various aspects of the overall project.

In many respects, effective farm management is similar to the management processes that are employed with any type of business. There are decisions that must be made on a daily basis, as well as operational guidelines that must be observed by everyone who is involved with the operation. Some participants are accountable to overseers or managers, who in turn are accountable to owners.

What sets farm management apart from other moneymaking ventures is the kinds of daily duties involved and the number of management layers found in the operation. Even among farms, the processes will vary depending on the type of farming business involved and the overall size of the operation. For example, the specific tasks associated with effective dairy management will be somewhat different than tasks connected with the operation of a wheat farm.

Large commercial farms are more likely to rely on modern technology in order to maximize the efficiency of the management process. Special software that makes it possible to keep track of units produced, units in process, pending orders, and outstanding expenses help to simplify a number of management tasks. In addition to this software, state of the art equipment for cultivation, harvesting, and maintenance also help enhance the productivity of farms of all sizes.

Even small farmers can benefit from management advice. Consulting with a farm management company makes it possible for experts to assess the current condition of the farm and make constructive suggestions on how to increase the productivity and profit margin for the farm. Managers assigned by the company can oversee the implementation of the suggestions while training family members and hired help how to use the additions to best advantage.

As with many career options, training in the field normally involves a combination of structured study in a degree program provided by a college or university and on the job experience. Some degree programs require that students work as interns at commercial farms before receiving their educational credentials. Even long time farmers can study management basics through the use of materials available from various farming associations and agencies, however.

1.3 MEANING OF FARM MANAGEMENT

Farm Management comprises of two words i.e. Farm and Management. Farm means a piece of land where crops and livestock enterprises are taken up under common management and has specific boundaries.

Farm is a socio economic unit which not only provides income to a farmer but also a source of happiness to him and his family. It is also a decision making unit where the farmer has many alternatives for his resources in the production of crops and livestock enterprises and their disposal. Hence, the farms are the micro units of vital importance which represents centre of dynamic decision making in regard to guiding the farm resources in the production process.

The welfare of a nation depends upon happenings in the organisation in each farm unit. It is clear that agricultural production of a country is the sum of the contributions of the individual farm units and the development of agriculture means the development of millions of individual farms.

Management is the art of getting work done out of others working in a group.

Management is the process of designing and maintaining an environment in which individuals working together in groups accomplish selected aims.

Management is the key ingredient. The manager makes or breaks a business.

Management takes on a new dimension and importance in agriculture which is mechanised, uses many technological innovations, and operates with large amounts of borrowed capital.

The prosperity of any country depends upon the prosperity of farmers, which in turn depends upon the rational allocation of resources among various uses and adoption improved technology. Human race depends more on farm products for their existence than anything else since food, clothing – the prime necessaries are products of farming industry. Even for industrial prosperity, farming industry forms the basic infrastructure. Thus the study farm management has got prime importance in any economy particularly on agrarian economy.

1.4 DEFINITIONS

1. Farm management is that branch of agricultural economics which deals with the business principles end practices of farming with an object of obtaining the maximum possible return from the farm as a unit under a sound farming programme.

2. Farm Management may be called a science of decision making. Therefore it can be defined as a science which deals with judicious decisions o the use of scarce farm resources, having alternative uses to obtain maximum profit and family satisfaction on a continuous basis from the farm as a wholes.

3. Warren: Farm management is the study of the business principles of forming. It may be defined as the science of organization and the management of the farm enterprise for the purpose of securing the greatest continuous profits.

1.5 NATURE OF FARM MANAGEMENT

Farm management deals with the business principles of farming from the point of view of an individual farm. Its field of study is limited to the individual farm as a unit and it is interested in maximum possible returns to the individual farmer. It applies the local knowledge as well as scientific finding to the individual farm business.

Farm management in short be called as a science of choice or decision making.

1.5.1 Objective

Main object of farm management is to obtain the maximum net profit from the various enterprises on a farm. The main aim is to get maximum net returns from the farm as a whole. This leads to success. This object constitutes selection, combination and execution of enterprises consistent with a sound agricultural policy.

The farm management study is undertaken with the following objectives:

1. To study the input output relationship in agriculture and determine the relative efficiency of various factor combinations.

2. To determine the most profitable crop production and livestock raising methods.

3. To study the cost per hectare and per quintal.

4. To evaluate the farm resources arid land use.

5. To study the comparative economics of different enterprises.

6. To determine the relation of size of farm to land utilization, cropping pattern, capital investment and labour employment.

7. To study the impact of technological changes on farm business.

8. To find out ways and means for increasing the efficiency of farm business through better input-output relationship and proper allocation of resources among different uses.

1.6 IMPORTANCE OF FARM MANAGEMENT

Proper management of farmland is vital for an investor to capitalize on the overall appreciation of the asset. Farming today is more than just producing crops, it requires farmers and landowners to address profitability, fertility, conservation, and tax issues to name just a few. The importance of a knowledgeable and professional farm manager is essential for maximizing the appreciation and income of investment farmland.

All farmland is not created equal and a customized farm management plan and oversight will align the interests of the farmer and landowner to optimize their return on investment (ROI). The key to proper farm management includes focusing on the following areas:

- Profitability
- Leasing
- Production
- Fertility
- Conservation
- Capital Improvements

- ◉ Additional Revenue Opportunities

- ◉ Insurance

- ◉ Taxes

- ◉ Communication

Overlooking just one of these key tasks can lead to a significant loss in the ROI or degradation of the farmland.

Farming over the last decade has become one of the most profitable industries, although the improvement in economics has not necessarily flowed back to the landowners. We estimate that on average, landowners across the U.S. are only receiving 50% of their potential rental income. This means that U.S. farmland owners are leaving $25 billion of rental income on the table.

Professional farm management services will not only allow investors to optimize their ROI, but own an asset that can be passed down for generations.

1.6.1 Profitability

Productive cropland is profitable on multiple levels by producing food to the growing population and also providing its owner with intermittent cash flow with stable appreciation upside. U.S. cropland has returned its investors 10.6% annually over the past 14 years via appreciation and rental contracts according to the USDA. Strong global demand for commodities grown in the U.S., including corn, soybeans, and wheat, has positioned U.S. farmland to continue to appreciate while returning annual rental rates of over 5% of the land's value. Farmland deserves to be a part of every well diversified investor's portfolio.

Farm management is essential for farmland owners to maximize annual ROI and long-term capital appreciation. Any farmland should increase in value and produce annual income to land owners, but with progressive farm management, landowners can expect much higher profitability.

1.6.2 Leasing

A key part of the farm manager's duties are their relations with the tenant operator. Choosing the appropriate operator can make or break an investment in farmland. The operator will help determine the short and long-term fertility, production, conservation, and cosmetics of a property. Even one mismanaged year of farming can cause significant damage to a property.

To ensure the right tenant is chosen, managers will interview many qualified farmers, including a thorough inspection of the tenant's operation. Background checks are also important as lenders and local contacts will give a better insight to the credit worthiness of the farmer.

The manager should have a large pool of potential tenants that are competing for a given property. Competition between operators for leasing a property will give the landlord the highest possible rent. A good manager will know what market rent is in the area and use that as a starting point for negotiating.

Identifying the right type of lease will differentiate a good farm manager from a great farm manager. Each lease presents different cash flow and risks. It is imperative that the owner and manager are on the same page and comfortable with the chosen lease and the amount of risk exposure to the landlord.

There are Four Commonly Used Leases:

- ◉ Cash Rent Lease: Tenant pays a specified amount of cash per acre per year to farm the property. The payment is made in full before any seed is planted; typically paid on March 1st of the leased year.

- ◉ Flex Lease: This lease is a variation of the Fixed Cash Lease. Rent will vary based on yields and crop prices throughout the year. This lease gives the landlord a share of the risks associated with farming in a given year.

- ◉ Crop-Share Lease: Tenant pays landlord with a percent of the crop or income that is produced. In a typical lease, the landlord would split the input farming costs of fertilizer, seed, pesticides, etc. In some instances, the landlord is responsible for marketing their share of the crop as well.

- ◉ Custom Farm Lease: This is the most involved an owner can be in a farming operation without actually farming. The owner takes on the input costs including fertilizer, seed, pesticides, herbicides, etc. while an operator is contracted out by the owner for performing farming tasks throughout the year. 100% of the production is retained by the owner.

Although each lease has its differences, there are two important aspects in common; price and length.

Price of rent can be the deciding factor on whether or not to purchase a property for investment. A good manager will not only use market averages, but also forecast an income statement for the property to estimate what the owner's share of income should be.

The second part of negotiating a lease is the length. A long-term lease is not always in the landlord's best interest. If a lease is signed for $100 per acre for five years, and after two years the market rent increased from $100 to $125, the landlord is missing out on $25 per acre. Negotiating a one to three year lease ensures the landlord will be receiving the current fair market rent.

1.7 PRODUCTION

It is always important to track crop conditions, but for certain leases, including flex and custom farming where the landlord has upside potential in the production of the property, crop conditions are of utmost importance. Farm managers work with farmers to ensure planting was successful and the correct seed varieties were planted for the climate forecasted during the upcoming growing period.

Throughout the planting season, farm managers keep in contact with operators to note how the crops are progressing which will help build a strong historical file for the property. Farmland with consistent proven yields of 200 bushels of corn will have a higher value than a property that has a volatile production history of similar soil quality as future buyers prefer consistent yielding properties.

Harvest will produce yield data that farm managers record for the property's historical file. Often operators will have a yield map that will be supplied to the farm manager, helping the manager understand where the strongest yielding areas of the field are located along with other features like compaction or poor drainage. Matching soil types from a soil map to the yield map often will reveal where the poorly drained areas of the property are located. A manager will then explore where additional drainage relief is needed, let it be tile, surface intakes, or a waterway.

1.7.1 Fertility

Once harvest is complete and farmers begin to plan their input purchases for the following year, farm managers will work with farmers to gauge the fertility of the property with the use of soil samples. Farmers test soil fertility via soil samples at least every other year to make sure the correct amount of fertilizer is used to achieve optimum yields.

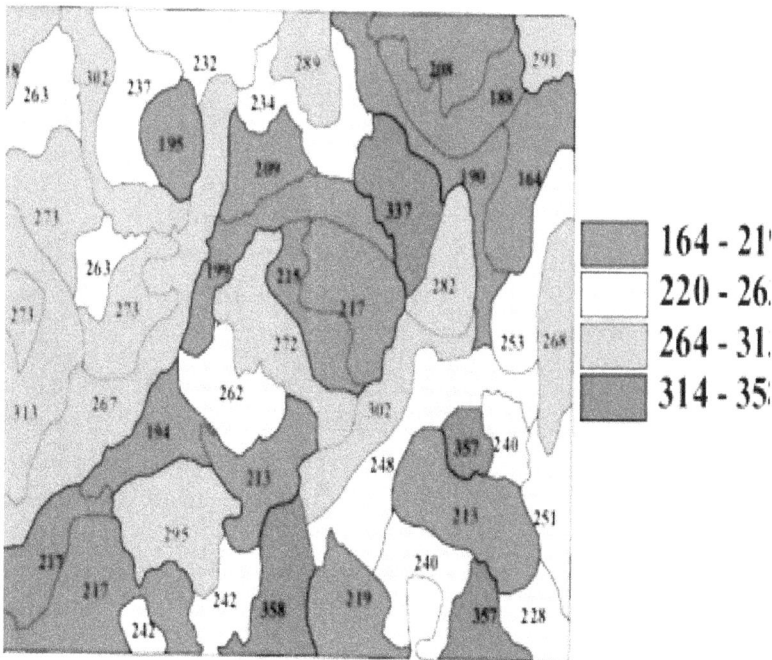

Fig. 1. Potassium soil sample in parts per million.

Additionally, farmers do not want to apply too much fertilizer than their soil Cation-Exchange Capability (CEC) can handle which would lead to wasted fertilizer and money. Soil CEC refers to the amount of nutrients a soil can absorb efficiently at a given pH level. If a soil has a low CEC, then over fertilizing can lead to fertilizer runoff and waste.

Farm managers work with the farm operator after comparing yields maps, soil maps, and soil sample maps to discuss the property's nutrient program moving forward to meet fertility goals while maximizing yields.

Tracking soil fertility through soil samples is essential for future property value increases. Typically, purchasers do not want to buy a property with poor nutrient levels. Although poor fertility is not typically permanent, application of macro and micro nutrients along with lime are required to bring fertility up to adequate levels over a period of multiple years which can carry significantly high costs and lead to lower rents. If a landowner currently holds a property with poor fertility, it is important to have a farm manager work with the operator to build a nutrient program to rebuild the property's fertility.

1.8 CONSERVATION MANAGEMENT

Farmland is arguably the most important asset to sustaining life on an ever growing planet. World population is growing at an exponential rate, and taking care of and conserving farmland is essential to feeding the growing population. In order to conserve our precious asset, the USDA developed the Natural Resource Conservation Service (NRCS). The NRCS helps land owners reduce soil erosion, enhance water supplies, improve water quality, increase wildlife habitat, and reduce damages from floods and other natural disasters.

There are a variety of programs a farmland owner could sign up for; the programs most used by farmland owners include the Conservation Reserve Program (CRP), Grassland Reserve Program (GRP), Water Bank Program (WBP), Wetland Reserve Program (WRP), and Emergency Watershed Program (EWP). Each has its own unique characteristics, but in general when land is enrolled in these programs, the owner receives yearly payments and the contracts typically last 10 to 15 years.

In order to sign up for these programs, the farm manager must put in a considerable amount of work and due diligence. Having a good relationship with the local NRCS office can help simplify signing up for a conservation program. Depending on when a landowner starts to sign up for a program, communication with the NRCS office will last for one to two months. Weekly interactions via phone or e-mail are a must and having a good relationship with an NRCS representative is essential for a smooth process.

To start the process of signing up for a conservation program, a manager must first have detailed knowledge of the land and what areas they would like to have signed into a program. Aerial maps showing outlines of possible areas will need to be procured and presented to the NRCS. Upon review of maps and physical examination by the NRCS, they will determine what type of conservation program the land falls under. Each program has its own due dates for signing up. It is crucial that the work and due diligence be done in a timely manner as these sign ups typically only come once a year.

Upon approval of the conservation program, a manager will have to work with the current operator to make sure they meet all the requirements the NRCS has set forth. This may include seeding grass, plugging drains, spraying for weeds, or excavating. Getting these requirements completed correctly and in a timely manner is essential as the NRCS does periodic, onsite, evaluations. If the requirements are not met or part of the agreement is breached, it could result in fines or payment stoppage. This should not happen if the manager has done their due diligence and seen the process through to the end.

1.8.1 Adding Value Through Capital Improvements

In order to maximize appreciation, ideal farmability should be targeted by the farm manager and owner which will include capital improvement projects. By taking a diverse approach to capital improvements, farm managers can present projects that can fit any landowner's budget. Common capital improvements cover drainage, erosion, and access.

Adding drain tile is one of the single best additions one can make to a property. The primary reason to install drainage tile in a farm field is to increase productivity through healthier crops. Ideal soil is made up of 50% soil, 25% water, and 25% air. When a heavy rain elevates the water table, the soil loses its 25% air make-up, which will hurt crop growth and increase soil pH levels. Fixing drainage issues by installing drainage tile typically pays for itself through increased yields within five years of installation by farmers paying higher rent. Tile projects have a wide range of cost from small localized projects totaling $1,000 to $2,000 to large scale parallel pattern tile projects running upwards of $750 per acre.

The cost of adding drainage tile can often be immediately added on to a property's value. Since drainage tile is eligible for accelerated depreciation, farm managers work with accountants, farmers, excavators, and previous owners to assign a value to any tile in a property so the landowner can write off the cost as a 100% tax deduction. An accurate tile depreciation valuation can save landowners tens of thousands of dollars on taxable income. Additionally, working installation of drainage tile into a lease can provide opportunity for a landowner to get discounted tile if their operator installs the tile at a reduced rental price.

Precious soil can erode away via the wind, rain, or other weather elements thus striping a landowner of their asset. Farm managers are aware of erosion issues and should be constantly monitoring every property for signs of erosion so they may act fast to limit any soil loss. Simple solutions to localized erosion including installing a berm or retaining wall and extreme measures including installation of waterways, ditches, terraces, or other major excavation work which would be administered by the farm manager.

Fig. 2. Erosion before and after management.

Another excellent way to increase property value is by adding field access points. On a yield map managers will note poor yields that were caused by soil compaction. Often soil compaction is caused by heavy machinery running over the same area repetitively; often near field access points. By adding multiple points of access, farmers can cut down their traffic on compaction areas, thus increase total production over time.

In years of drought, any farmland would benefit from an irrigation system. Farm managers will work with irrigation outfitters to price an irrigation system and generate a long-term economic analysis of installing an irrigation system. Depending on the property's location and soil makeup, irrigation can substantially increase yields, cash rents, and position the property for ultimate appreciation.

1.8.2 Additional Revenue Opportunities

Revenue can be created on farmland, not just via farming operations, but also through other means including wind easements, hunting rights, and advertisements.

Landowners can benefit from having a wind farm a part of their property by leasing the property's wind rights. These contracts are created in the first process of building a wind farm, so land owners get paid prior to any building. When the wind farm is finalized and constructed, land owners will receive fixed and variable payments based on electricity production.

Landowners could receive up to $15,000 per year on a 160 acre parcel, although each wind company's contract will differ. Farm managers will handle agricultural impact and economic research behind any such wind project, keeping in mind the property's future for appreciation at all times.

Fig. 3. Additional revenue opportunities.

Midwest farmland produces the best corn yields in the world, but also some of the best hunting as well. Upland birds, waterfowl, and deer hunting are some of the most sought after hunting experiences that outdoorsmen demand. By leasing out farmland to hunters for hunting rights, landowners can generate increased annual ROI on top of their crop production lease. Farm managers will source tenants, work with an attorney to draft a proper lease, recommend insurance protection, and manage the hunting rights lease on farmland.

Often farmland is located on a desolate gravel road upwards of 15 miles from the nearest housing development or town, but sometimes farmland will be located on a busy highway or interstate with high amounts of traffic. Roadways with high traffic levels lead way for advertisement potential on neighboring farmland. Progressive farm managers will investigate advertisement potential of property and work with a billboard company to put together an investment proposal to see if billboard advertisement would generate additional ROI via rent payments and increased land value.

Insurance

A good farm manager will go above and beyond standard management to insure a client has protection against liability. Farming can be a dangerous job with many hazards as large machinery will be used on a given property throughout the year. Accidents can happen, and making sure a client is protected from liability is important. Typically a farm manager will not administer insurance, but pointing the client in the right direction and helping with the process is a duty a manager should embrace.

Taxes

Property taxes are paid each year, due on specific dates that vary by state. A manager will need to be familiar with each state's due dates to ensure their client is compliant as clients might have multiple properties across several states. Failure to pay property taxes could result in fines on the property, so educating a client on property taxes is essential to a manager's duties.

Communication

Communicating with clients is vital throughout the management process to ensure the client and managers are in collaboration to meet shared goals. As described in detail in the paragraphs above, each management duty is conducted for a reason. Communicating why certain duties are performed will keep all parties on the same page.

Typically, the farm manager will provide a written annual report, highlighting the previous farming season, the outlook for the coming season, and potential capital improvement projects. This will assist the client in documenting the farming operation and better understanding the economics of their investment.

Essential Management

The unique investment opportunity farmland provides to investors requires hands on management unlike other asset classes. Farmland is exposed to not only economic elements, but the tangible weather elements as well, making oversight of the asset more important and strenuous. High-quality farm management will help add value and maximize ROI by providing proactive guidance and supervision to farmland property.

1.9 ROLE OF IT IN AGRICULTURE

The role of Information technology in agricultural sector is becoming more and more visible. We use IT to convey and spread information to people on matters relevant to crop production and crop protection. People must have a computer or a computing device like smartphone to avail use of information technology. Promulgation of information alone can't sustain growth in agriculture; agricultural industry must have the ability to manipulate that information to make informed decisions. I.e. in the agriculture context, decisions which will have a positive impact on related activities are being made. Modern farming practices like satellite farming has already started getting popularity in foreign countries, this precision farming uses IT to make direct contribution to maximize the crops productivity. Satellite technology, geographic information systems, remote sensing, techniques of agronomy and soil science etc. can be used to increase agricultural production especially in large tracts of land where this approach is cost effective and useful. Big food chain retailers has started implementing this competing technological aide in their vast lands.

However in Indian context the potential of IT in farming sector seems unexploited. Lack of awareness about the technologies is a major constraint. Currently the farmers in India hesitate to come out of the entanglements of traditional source of inputs. Using information is not only useful but a requirement in these days. Studies shows that the use of modern technologies is capable of making remarkable hike in agricultural production.

1.9.1 Needs of the Hour

Decision Support System (DSS) for Farmers

Farmers should be cautious in decision making which help them to avoid impending risks. The refined exporting rules proclaimed by WTO will make exports more competitive. Harvesting costs effective farming methods and the availability of data inputs against imports will facilitate the assessment of the market value of the indigenous products. Analyzing these kind of data will help the farmers to make necessary steps and corrective measures to face the market situations.

Market Monitor

Fluctuations in the international market will directly affect the markets too, so it is necessary to be sharp-eyed to defend external shocks. The market watch is that important. Advance warning systems should be developed which would help farmers to make last minute strategy changes to avoid huge loss. Although periodic analytical reports are needed to enhance the warnings.

Opportunities

Indian farmers should seize the opportunities. They should equip their business with every possible technologies to cop up with the business. The whole agricultural sector is in the cusp of an impending revolution, thankfully central and state government departments are well aware about the scenario. Thus a wide range of awareness campaigns are on the go.

1.10 SIGNIFICANCE OF AWARENESS

In India farming with agricultural technologies is an unprecedented practice, so the foremost consideration of WTO should be given to unambiguous interpretation and implication of training methods. It is recommended that WTO should give an effort to publish printed, web journals targeting all segments of Indian agriculture and allied activities. This has to be addressed immediately as a priority item. The most sought after information like import tariffs, season wise and year wise phases of the mandatory changes in government policies and its impact on the various subsidy schemes are to be furnished to the beneficiaries. How to provide this analytical inputs will lead us to the advantage implementation of web/mobile applications for agriculture.

Removing restrictions to throw open the Indian agricultural markets, macro-economic situation, foreign exchange, inflation, the current tariff, etc. in the respective countries are likely to have a direct impact on Indian agriculture segments inside and outside.

1.11 FARMING IN INDIA

The Indian farmer had discovered and begun farming many spices and sugarcane more than 2500 years ago. Did you know that our country is the 2nd largest producer of agricultural products in the world? In fact, agriculture contributes as much as 6.1% (as of 2017) to our Gross Domestic Product (GDP).

1.11.1 Agricultural Methods of the Indian Farmer

Farming is one of the oldest economic activity of our country. Different regions have different methods of farming. However, all these methods have significantly evolved over the years with changes in weather and climatic conditions, technological innovations and socio-cultural practices.

Primitive Subsistence Farming

This is a primitive farming method and farmers still practice it in some parts of the country. While this type of subsistence farming is typically done on small areas of land, it also uses indigenous tools like a hoe, Dao, digging sticks etc. Usually, a family or the local community of Indian farmers are engaged in this farming method who use the output for their own consumption. This is the most natural method, where the growth of crops but dependent on the rain, heat, fertility of the soil and other environmental conditions.

The key to this farming technique is the 'slash and burn' method. In this practice, once the crops are grown and harvested, the farmers burn the land. They then move to a clear patch of land for a new batch of cultivation. As a result, the land gains back its fertility, naturally. Because no fertilizers are used for cultivation, the primitive subsistence method yields good quality crops and also retains the properties of the soil.

Plantation Agriculture

Plantation agriculture was introduced in India by the Britishers in the 19th century. This type of agriculture involves growing and processing of a single cash crop purely meant for sale. Large capital input, vast estates, managerial ability, technical know-how, sophisticated farm machinery, fertilizers, good transport facilities, and a factory for processing the produce are some of the outstanding features of plantation agriculture.

There are plantations of rubber, tea, coffee, cocoa, banana, spices, coconut, etc. This type of agriculture is practised mainly in Assam, sub-Himalayan West Bengal, and in the Nilgiri, Anaimalai and Cardamom Hills in the south.

Shifting Agriculture

This is a type of agriculture in which a piece of forest land is cleared mainly by tribal people by felling and burning of trees and crops are grown. After 2-3 years when the fertility of the soil in the cleared land decreases, it is abandoned and the tribe shifts to some other piece of land.

The process continues and the farmers again shift to the first piece of land after a gap of 10-15 years. This type of agriculture is practised over an area of 54 lakh hectares, 20 lakh hectares being cleared every year. Dry paddy, buck wheat, maize, small millets, tobacco and sugarcane are the main crops grown under this type of agriculture.

This is a very crude and primitive method of cultivation which results in large scale deforestation and soil erosion especially on the hill sides causing devastating floods in the plains below. About one million hectares of land is degraded every year due to shifting agriculture.

Intensive Farming

In areas where irrigation has been possible, the farmers use fertilisers and pesticides on large scale. They have also brought their land under high yielding variety of seeds. They have mechanised agriculture by introducing machines in various processes of farming.

Also known as industrial agriculture, it is characterized by a low fallow ratio and higher use of inputs such as capital and labour per unit land area. This is in contrast to traditional agriculture in which the inputs per unit land are lower.

1.11.2 Remember Intensive Agriculture Development Program

Intensive Agriculture Development program (IADP) was the first major experiment of Indian government in the field of agriculture and it was also known as a "package programme" as it was based upon the package approach.

The programme was launched in 1961 after the Community Development Programme lost sheen. The core philosophy was to provide loan for seeds and fertilizers to farmers. Intensive Agriculture Development program was started with the assistance of Ford Foundation.

The IADP was expanded and later a new Intensive Agriculture Area programme (IAAP) was launched to develop special harvest in agriculture area.

Crop Rotation

This refers to growing of number of Crops one after the other in a fixed rotation to maintain the fertility of the soil. The rotation of crops may be complete in a year in some of the areas while it may involve more than one year's time is others.

Pulses or any leguminous crop is grown after the cereal crops. Legumes have the ability of fixing nitrogen to the soil. Highly fertilizer intensive crops like sugarcane or tobacco are rotated with cereal crops. The selection of crops for rotation depends upon the local soil conditions and the experience and the understanding of the farmers.

1.12 SEDENTARY CULTIVATION OR PERMANENT AGRICUL-TURE

It is also known as settled cultivation. In it farmers get settled at the place and practice continued use of land year after year with the variation of crops. In it permanent settlement of farmers exists. It is the normal system of agricultural practice found in almost every part of India.

1.12.1 Terrace Cultivation

Where lands are of sloping nature, this type of cultivation is practiced specially in hilly areas. The hill and mountain slopes are cut to form terrace sand the land is used in the same way as in permanent agriculture. Since the availability of flat land is limited terraces are made to provide small patch of level land. Soil erosion is also checked due to terrace formation on hill slops.

1.12.2 Dry Farming

Farming which is totally based on rainfall is called dry farming. This is a type of traditional farming. Such farming is found in low rainfall region and where irrigation facility is not available. Some parts of the states like Rajasthan, Gujarat, Maharashtra, Karnataka, Tamil Nadu, Andhra Pradesh, Madhya Pradesh, etc are practice dry farming. Rain water percolates in the land, such moist land is utilised for agriculture.

1.12.3 Features

⦿ Rain water percolates in the soil. It is used for growing Kharif and rabi crops. Food crops are mostly grown in this type of farming. Jowar, bajra, corn, moong, groundnuts etc. are grown in Kharif season. In the rabi season crops like wheat, gram, soyabean, sunflower, linseed, etc are cultivated.

Mixed Cropping

Mixed cropping means the cultivation of more than one crop simultaneously on the same piece of land. The two crops are sown together but harvested at different times because the growth period of the plants of the different crop varies. Early maturing crops are mixed with groundnut, cotton or pulses which mature late. The crops are so mixed that soil nutrients removed by some are replaced by others, at least partly.

Arable Farming

It is a system under which the farms are used only for the cultivation of crops i.e. food crops and cash crops. Mixed farming is a system under which the lands used not only for the cultivation of crops, but also for other purposes, such as stock-raising, poultry farming, sericulture, bee-keeping etc.

1.12.4 Types of Plants Grown

We can also divide the crops of India in four segments on the basis of crops as follows:

Food Grains

This includes Rice, Wheat, Maize, Coarse Cereals and Pulses

Cash Crops

This includes Cotton, Jute, Sugarcane, Tobacco and Oilseeds. Oil seeds include ground nut, Rapeseed & Mustard, Sun-flower, Soyabean etc.

Plantation Crops

This includes Tea, Coffee, Coconut, Rubber etc.

Horticulture

This includes Fruits and Vegetables.

1.12.5 Types of Seasons Crops

Further, the crops of India can also be divided in three types on the basis of their seasons viz. Rabi, Kharif and Zaid.

Kharif Crops

The Kharif crop is the summer crop or monsoon crop; usually sown with the beginning of the first rains in July, during the south-west monsoon season. Major Kharif crops of India include Millets (Bajra and Jowar), Paddy (Rice), Maize, Groundnut, Red Chillies, Cotton, Soyabean, Sugarcane, Turmeric etc.

Rabi Crops

Rabi crop is the spring harvest or winter; sown in October last and harvested in March April every year. Major Rabi crops in India include Wheat, Barley, Mustard, Sesame, Peas etc.

Zaid Crops

Zaid is grown in some parts of country during March to June. Prominent examples are Muskmelon, Watermelon, Vegetables of cucurbitacae family such as bitter gourd, pumpkin, ridged gourd etc.

1.12.6 Farmer Education

Farmers require ongoing education to stay aware of fast-moving developments in technology, science, business management, and an array of other skills and fields that affect agricultural operations. NIFA initiatives increase farmers' knowledge in these areas and help them adopt practices that are profitable, environmentally sound, and contribute to quality of life.

1.12.7 Importance of Farmer Education

Farmers — beginning and experienced — are critical to creating rural prosperity in the United States. However, farmers face unique challenges and require education and training to ensure their success.

Training helps farmers to incorporate the latest scientific advances and technology tools into their daily operations. The results of enhancing their operations with these tools increases efficiency and can also lead to:

- Less harm to the environment
- Reduced food contamination
- Reduction of the need for water and chemicals for crops
- Increased profits

1.13 ECONOMIC PRINCIPLES APPLIED TO FARM MANAGE-MENT

The outpouring of new technological information is making the farm problems increasingly challenging and providing attractive opportunities for maximising profits. Hence, the application of economic principles to farming is essential for the successful management of the farm business.

Some of the economic principles that help in rational farm management decisions are:

1. Law of variable proportions or Law of diminishing returns: It solves the problems of how much to produce ? It guides in the determination of optimum input to use and optimum output to produce.. It explains the one of the basic production relationships viz., factor-product relationship

2. Cost Principle: It explains how losses can be minimized during the periods of price adversity.

3. Principle of factor substitution: It solves the problem of 'how to produce?. It guides in the determination of least cost combinations of resources. It explains facot-factor relationship.

4. Principle of product substitution: It solves the problem of 'what to produce?'. It guides in the determination of optimum combination of enterprises (products). It explains Product-product relationship.

5. Principle of equi-marginal returns: It guides in the allocation of resources under conditions of scarcity.

6. Time comparison principle: It guides in making investment decisions.

Principle of comparative advantage: It explains regional specialisation in the production of commodities.

1.14 RELATIONSHIP OF FARM MANAGEMENT WITH OTHER SCIENCES

The Farm Management integrates and synthesises diverse piece of information from physical and biological sciences of agriculture.

The physical and biological sciences like Agronomy, animal husbandry, soil science, horticulture, plant breeding, agricultural engineering provide input-output relationships in their respective areas in physical terms i.e. they define production possibilities within which various choices can be made. Such information is helpful to the farm management in dealing with the problems of production efficiency.

Farm Management as a subject matter is the application of business principles n farming from the point view of an individual farmer. It is a specialised branch of wider field of economics. The tools and techniques for farm management are supplied by general economic theory. The law of variable proportion, principle of factor substitution, principle of product substitution are all instances of tools of economic theory used in farm management analysis.

Statistics is another science that has been used extensively by the agricultural economist. This science is helpful in providing methods and procedures by which data regarding specific farm problems can be collected, analysed and evaluated.

Psychology provides information of human motivations and attitudes, attitude towards risks depends on the psychological aspects of decision maker.

Sometimes philosophy and religion forbid the farmers to grow certain enterprises, though they are highly profitable. For example, islam prohibits muslim farmer to take up piggery while Hinduism prohibits beef production.

The various pieces of legislation and actions of government affect the production decisions of the farmer such as ceiling on land, support prices, food zones etc.

The physical sciences specify what can be produced; conomics specify how resources should be used, while sociology, psychology, political sciences etc. specify the limitations which are placed on choice, through laws, customs etc.

2

PRINCIPLES OF FARM MANAGEMENT

2 .1 OBJECTIVES

After studying this chapter you will be able to understand:

- Farm Management Principles.

- Poultry Farm Management.

- Management Procedure in Various Farm Management.

- Economic Importance of Fisheries.

2.2 INTRODUCTION

The term 'farm management' is used by different people to convey different concepts. Aquaculturists often tend to consider it as the overall technical operation of the farm and supervision of day-to-day activities. Good farm management expertise is often considered to be the same as practical experience in the application of aquaculture technologies in the field. Proper and timely maintenance of the farm and its installations, successful methods of brood stock manipulation, breeding, seed production, stocking, feeding, disease and pest control, proper water management, including the maintenance of water quality, protection of the stock from poaching, harvesting and marketing are the major elements of this concept of management.

The science of farm management, which is relatively new and developed in agriculture and animal production, is based on the concept of a farm as a business and consists of the application of scientific laws and principles to the conduct of farm activities. Originating in production or agricultural economics, it is now accepted as multidisciplinary science. Yang (1965) defined it as 'a science which deals with the proper combination and operation of production factors, including land, labour and capital, and the choice of crop and livestock enterprises to bring about a maximum and continuous return to the most elementary operation units of farming'. He considered it a pure science because it deals with the collection, analysis and explanation of facts and the discovery of principles; and at the same time an applied science because the ascertainment and solution of farm problems are within its scope.

Farm management involves a continuous process of economizing and therefore the relevant basic theory of farm management is economics. However, it has to draw heavily on biology, technology, meteorology, sociology, psychology and related disciplines to optimize the use of scarce resources. While scientific research to develop technologies is performed in laboratories and experimental stations, farm management research is done in the field by collecting and analysing information from individual farms, to discover or verify successful farming practices under specified circumstances. Its aim is to plan optimum farm organization and management practices for higher production efficiency and maximum farm earnings. Although field experimentation can be a useful means of generating the necessary information, cost considerations often militate against its use in developing farm management methods. Experiments have admittedly the advantage of elucidating clearly input/output relationships, by varying the level of selected inputs. By replication and statistical methods, the reliability of the results and significance of the differences can be measured. However, it usually fails to discover the interactive effects between factors. Because of this, many of the data required for management research are obtained through farm management surveys, financial bookkeeping and the study of farm practices, including costs, use of land, water, labour and other material requirements. The results of management research can be used by farm managers in planning their activities or by governments in formulating farm policies.

The main elements of the economic principles considered in agricultural farm management are comparative advantage, diminishing returns, substitution, cost analysis, opportunity cost, enterprise choice and goal trade-off.

Comparative advantage relates to the determination of the most economically suitable crops for a farm or area, from among the different crops that could be grown there. Since the comparative advantage can change as a result of changes in technology, input and transportation costs, farm product prices, etc., it will be necessary to evaluate the advantages on a continuing basis.

The principle of diminishing physical and economic returns determines the best level for any production practice. It helps in considering the level of output produced from a set of fixed resources, taking into account the variable factors. Diminishing economic returns appear when diminishing physical returns are converted into value, generally measured in money terms. For example, in considering the use of weedicides in a farm, the farmer has to balance the money cost involved against the expected money value of the increased yield, or losses prevented, in order to decide whether it pays from the financial standpoint. It may be that he should use the weedicide up to the point where the last unit of application is expected to pay for itself.

The principle of substitution refers to the selection by the farmer of the most economical method, measured in the most appropriate terms (e.g. physical labour, time or money) to suit his conditions. For example, the farmer has the option to use manual labour, mechanical equipment or chemical means to control weeds in his fish pond. He has to decide which of the methods he should use, taking into account the performance and cost of each. In substituting one method for another, he has to ensure that the saving is greater than the cost of the technique added. This principle is of special importance when decisions have to be made on the adoption of new practices.

The principles of cost analysis have been dealt. Even though the farmer may have some control over the costs of production on his farm, he has little control over the prices he receives for his produce. It is obvious that, under normal circumstances, a farmer must reduce his costs per unit of output if he is to increase his net farm income. While the fixed costs remain the same regardless of how much he produces, variable costs change as the size of operation changes. The classification of a particular cost as fixed or variable depends partly on the nature and timing of the management decision considered. For example, land rent becomes a variable cost in relation to a decision to lease more land; but for land already leased and being used, the rent is a fixed cost.

The importance of opportunity cost in farm planning and decision-making are indicated. This concept relates to the cost of any choice in relation to the value of the best alternative foregone. For example, if a farmer can earn a profit of $1500 from a farm growing milkfish and $3000 by growing shrimps, the opportunity cost of growing milkfish is $3000. If he persists in growing milkfish he should recognize that he is earning $1500 less profit than he could have earned. Although for certain reasons he may continue with milkfish, the general principle is that the land, water, labour and capital should be used where they will add most to the income. The income may be measured directly as money, or in some broader terms such as satisfaction or utility.

Enterprise choice is made by a farmer, making allowances for the relationship with other activities or enterprises on his farm. Enterprises can be supplementary or complementary, as in aquaculture integrated with crop and animal farming. As far as operation of his farm is concerned, the overall goal of the farmer is to make the most efficient use of whatever resources he has.

All the considerations summarized above relate largely to internal allocation of resources to enterprises and activities that will maximize the net return. The principle of goal trade-off implies the existence of multiple goals which will often compete with one another, such as cash income, utilization of unproductive land, export earnings, etc. The farm may be managed to achieve that mix of goal attainments which gives the farmer the best level of overall satisfaction across his multiple goals. There may have to be some trade-off, ensuring that the gain in satisfaction from the relatively more important goal is greater than the decreased satisfaction incurred on the other goals.

Application of the above economic principles in farm management is very much influenced by two factors that are somewhat unique to the farming of animals and plants, whether aquatic or terrestrial. One is the varying degree of uncertainty under which annual operations have to be planned. The uncertainty may refer to the climate that will prevail, natural disasters that can occur, incidence of pests and diseases, the performance of new technologies adopted, the prices and competition that may be faced in the markets and the political environment in which the enterprises have to operate. Decisions are made under such uncertainties and therefore call for the exercise of personal judgement by the manager about the risks that he faces in the application of the various economic principles. Conclusions based on historic data can be of only partial assistance and decisions have to be on the basis of estimated future possible yields, costs, prices and technology. The other important factor is the orientation of the farm: whether it is completely market-oriented and operating commercially in a money economy, or whether it is subsistence or semi-subsistence farming. A good majority of small-scale farmers, and almost all large-scale farmers, have contact with markets through which they receive money as total or partial income.

It may appear that the economic theories mentioned above do not apply to small-scale farmers who operate outside the cash economy. But in point of fact, they are very pertinent to their operations and can be used to assess the gains and losses, irrespective of whether money value or some other measure (such as utility or satisfaction) is used. When applicable, money is a very convenient measure, as it enables comparisons between farms and the aggregation of individual farm performance to regional and national aggregates. When gains and losses involve both cash and non-cash elements, the trade-off or exchange rates between them will be specific to each farm. So it has to be recognized that money is a compromise measure. While it may be the best basis for analysis, it is less than an adequate approximation, depending on the extent to which trading guides are available on the money value of non-cash gains and losses.

2.3 FARM MANAGEMENT PRINCIPLES

Farm management principles guide the farmer or farm manager to take decisions. Details on six basic principles involved in making rational farm management decisions are presented.

a. Principle of variable proportions or laws of returns
b. Cost principle
c. Principle of substitution between inputs
d. Equi-marginal returns principle or opportunity cost principle
e. Principle of substitution between products
f. Principle underlying decisions involving time and uncertainty.

2.3.1 Principle of Variable Proportions or Laws of Returns

⊙ This principle helps in deciding the optimum amount of an input that needs to be applied for cultivation of a particular crop or enterprise. In agriculture law of diminishing returns will operate.

Diminishing Returns

⊙ "If increasing amounts of one input is added to a production process while all other inputs are held constant, the amount of output added per unit of variable input will eventually start decreasing"

2.3.2 Cost Principle

⊙ Most of the producers give considerable importance to the cost of production while taking production decisions.

Accounting Periods

⊙ There can be two accounting or planning periods: short-run and long run. Short run is a period of time wherein at least some factors (such as land, buildings,) used in crop cultivation is fixed while others are variable and could be altered to increase the yield. The long run is generally considered to be the period wherein all the factors used for crop cultivation could be varied.

Application of the Fixed and Variable Cost Principle

⊙ In the short run, gross return must cover the variable costs. The maximum net revenue is obtained when marginal cost (MC) equals the price of the product (MR).

⊙ If gross returns are less than total costs (variable + fixed costs) but are still larger than the variable costs, guiding principle should be to keep increasing production as long as added returns (MR) are greater than added costs (MC).

⊙ In the long run, gross return should be more than variable plus fixed costs (total costs).

2.3.3 Principle of Factor – Substitution (Least-Cost Combination)

⊙ In agriculture, various inputs or practices can be substituted in varying degrees for producing a given output. A farmer can meet the nutrient requirement of the crop by applying, farm yard manure, vermicompost, neem coated urea, and other inorganic fertilizers. The inputs should be substitutable. The choice before the farmer is to either use only one particular source (organic / inorganic) to meet the entire nutrient requirement for the crop or choose a combination of sources. It is prudent for the farmer to choose a particular combination of inputs which would be of least cost to produce a given level of output. Cost minimization will not depend only upon the cost of inputs and prices of products but also on the rate of substitution.

2.3.4 The Law of Diminishing Marginal Returns

This law states that "An increase in the capital and labour applied to the cultivation of land causes in general a loss than the proportionate increase in the amount of produce raised unless it happens to coincide with an improvement in the art of agriculture."

There are three stages of the law of diminishing returns. They are 1) stage II and 3) stage III. The positions of the parameters i.e. TP (Total Product), AP (Average Product) MP (Marginal Product) and EP (Elasticity stages of production (see production function curve) are as under.

Stage-I (Irrational Zone)

1. This stage starts from origin and ends where AF & MF curves intersect each other.
2. The TP is increasing at increasing rate at first then at decreasing rate.
3. PP and MP both increase but MP is greater than IP.
4. The EP is greater than 1 (one)

Stage–II (Rational Zone)

1. It starts where PP & MP intersect each other and EP = 1. It ends when MP = 0
2. TP increases but at decreasing rate.
3. MP starts to decline continuously and AP also starts to decline but it is greater than MP
4. The elasticity of production (EP) is greater than zero but less than 1.

Stage-III

1. This stage starts when MP is zero and TP is at maximum.
2. TP starts to decline and it declines continuously.
3. MP becomes negative, remains positive.
4. EP is always less than zero.

2.3.5 Law of Equimarginal Returns

The farmer has only limited capital (including own and borrowed) at his disposal and he has to use this money among cultivation of crops and other enterprises that are technically feasible in his farm. The farmer has to make a choice of the crops to be cultivated and the allied enterprises to be undertaken and also the area under the selected crop and size of enterprise, such that it would maximize the net revenue from the farm as a whole. This principle, thus, states that resources should be used in crops or enterprises where they bring not the greatest average returns, but the greatest marginal returns.

The law of Equimarginal returns is concerned with the allocation of the limited amount of resource among different enterprises. The law states that "profits are maximized by using a resource in such a way that the marginal returns from that resource are equal in all cases."

2.3.6 Law of Substitution or Principle of Least Cost Combination

The objective of profit maximization can be achieved by two ways, one by increasing output and other by minimizing the cost. The minimization of cost can be possible by deciding the use of more than one resource in substitution of other resources.

The objective of factor-factor relationship is twofold:

Minimization of cost at a given level of Output.

2. Optimization of output to the fixed factors through alternative resource use combinations.

$$y = f(x_1, x_2, x_3, x_4 \dots \dots \dots xn)$$

Y is the function of x_1 and x_2 while other inputs are kept at constant. The relationship can be better explained by the principle of least cost combination.

A given level of output can be produced using many different combinations of two variable inputs. In choosing between the two completing resources, the saving m the resource replaced must be greater than the cost of resource added.

The principle of least cost combination states that if two factor inputs are considered for a given output the least cost combination will be such where their inverse price ratio is equal to their marginal rate of substitution.

Marginal Rate of Substitution

MRS is defined as the units of one input factor that can be substituted for a single unit of the other input factor. So MRS of x_2 for one unit of x_1 is

$$MRS = \frac{\text{Number of unit of replaced resource}(x_2)}{\text{Number of unit of added resource}(x_1)}$$

Price Ratio

$$PR = \frac{\text{Cost pce unit addcd resource}}{\text{Cost per unit of replaced resource}}$$

$$\frac{\text{Price of } X_1}{\text{Price of } X_2}$$

Therefore the least cost combination of two inputs can be obtained by equating MRS with inverse price ratio.

i.e. $X_2 * Px_2 = x_1 * Px_1$,

This combination can be obtained by following algebraic method or Graphic method:

A. Isoquant/Iso Product Curve:

Iso = equal and quant = quantity.

An Isoquant represents the different combinations of two variable inputs used in the production of a given amount of output.

Properties of Isoquant

1. They slope down ward to the right: If more of one is used less of another input will be employed at the given level of output.

2. They are convex to the origin.

3. Isoquant does not intersect: It is not possible to have different outputs from a single combination of inputs.

4. Slope of Isoquant represents the MRS.

Types

1. Convex Isoquant (decreasing rate) Good substitution

2. Straight line Isoquant → Perfect substitute

3. Right angel No substitution complement

2.3.7 ISO-Costline

An ISO-cost line indicates all possible combinations of two inputs which can be purchased with a given amount of investment fund (outlay).

Each combination of inputs has same total cost which includes the cost of two inputs. (X_1 and X_2) combined.

Total cost = $PX_1 * X_1 + PX_2 * x_2$,

Properties of ISO-Cost Line

1. As total outlay increases, the ISO-cost line moves higher and higher away from the origin and vis- a-visa.

2. The ISO-cost lines are straight.

3. Slope of ISO-cost line represents price ratio i.e. Px_1 / Px_2 when x_1 is taken on X axis and x_2 on y axis.

ISO-Cline

It is a line passes through the points of equal slope or MRS on an Isoquant surface. With the input price ratio being constant for each Isoquant the MRS between the inputs is the same for each level of output.

Ridge Line

These are also called as border line. Ridge lines join the end points of Isoquants. The area within the ridge lines is rational region of production arid beyond that the two regions are irrational. Therefore these lines represent the limits of economic relevance.

Expansion Path

All the least cost combination points are joined to each other; the result is an expansion line. As such, MRS = PR.

Ridge (Border) Line

Line joining the end points of Isoquant.

2.3.8 Principle of Combining Enterprise

This principle is very important as it describes the product – product relationship. Here, instead of considering the allocation of inputs among enterprises, we discuses enterprise combination or product mix involving product relationship. Algebraically the relationship can be written as under:

There can be various relationships that can exist between enterprises or products:

Joint Product

Two or more than two products are produced in the same production process Eg. Paddy and straw (Agricultural products).

Complementary Productions

In this case relationship is directly proportionate. With the increase in one product there is also increase in other product. E.g. the cultivation of leguminous crop followed by cereals gives this relationship.

Supplementary Productions

In this case, increase in one product does not effect for each other or they are independent and if relationship is there it is supplementary Eg. Crop production and dairy enterprise.

Competitive Relationships

Here two products are said to be competitive when increase one needed to be reduction in other product e.g. Two cereal crops

2.4 DETERMINATION OF OPTIMUM PRODUCTION COMBINATION BY GRAPHIC METHOD

2.4.1 ISO-Revenue Curve

It is the line which indicates the different combinations of two products which gives the same amount of revenue or income.

Properties of ISO-Revenue Line

It is always straight line because the output prices do not change with the quantity sold.

The position of ISO revenue line shows the magnitude of the total revenue. As total revenue increases, the line moves away from the origin and vis-a-visa.

The slope of ISO-revenue curve represents the price ratio of two competing products.

2.4.2 The Law of Opportunity Cost

1. The opportunity cost is also called as alternative cost.
2. Opportunity Cost is the earning from the next best alternative sacrificed.

2.4.3 Law of Comparative Advantage

1. The concept of comparative advantage is associated with:

A. Resource productivity and

B. Cost of production of enterprise.

2.5 MANAGEMENT PROCEDURES IN VARIOUS FARM SYSTEMS

2.5.1 Dairy Farm Management

It is the management of animals for milk and its products for human consumption. Dairying generally deals with the processes and systems to improve quality and quantity of milk. Milk yield mainly depends on the quality of breeds.

The dairy farm management includes following processes:

(i)   Selection of good breeds containing high yielding potential (under the climatic conditions of the area) and resistance to the diseases.

(ii) Cattle should be housed-well, have sufficient water and should be kept in diseased-free conditions.

(iii) They should be fed in scientific manner with the emphasis on quality and quantity of fodder.

(iv) Regular inspection and keeping proper records of all the activities of dairy is also important.

(v)   Regular visits of a veterinary doctor is necessary.

(vi)   Stringent cleanliness and hygiene of both the cattle and the handler are very important during milking, storage and transport of milk and its products.

2.5.2 Poultry Farm Management

The term 'poultry' means rearing of domesticated birds, i.e., fowls, geese, turkeys and some varieties of pigeons but more often is used for fowl rearing.

Fowls are reared for food or for their eggs. Poultry birds reared for meat are called broilers and layers are female fowls raised for egg production.

The poultry farm management includes following processes

(i)   Selection of disease-free and suitable poultry breeds.

(ii)   Housing should be safe and provided with proper ventilation.

(iii)   Proper food and water should be provided.

(iv)   Healthcare and hygiene of poultry birds is mandatory.

Animal Breeding: It is an important aspect of animal husbandry, which aims to increase the yield of animals and to improve the desirable qualities of produce.

Breed is a group of animals related by descent and similar in most characters like general appearance, features, size, configuration, etc.

Animal breeding can be classified as:

Inbreeding

The crossing of closely related animals within the same breed for 4-6 generation is called inbreeding.

The strategies for inbreeding are:

(i)   Superior males and females of the same breed are identified and then mated. A superior male is bull, which gives rise to superior progeny as compared to other males.

(ii)   The progeny obtained from such type of matings are evaluated and superior males and females among them are identified for further mating.

(iii)   In case of cattle, more milk per lactation is the criteria for superior female for cow and buffalo.

Following are the advantages of inbreeding:

(i)   Inbreeding is necessary to evolve a pure line in any animal breed.

(ii)   It exposes harmful recessive genes that are eliminated by selection.

(iii)   Superior genes can be accumulated and inferior or undesirable genes can be eliminated by inbreeding.

(iv) Productivity of inbreed population can be increased by the selection at every step.

(v)   It increases homozygosity (the state of possessing two identical forms of a particular gene, one inherited from each parent).

(vi) sContinued inbreeding (mainly close inbree¬ding) reduces fertility and even productivity. This is called inbreeding depression.

To overcome this inbreeding depression, related animals of the breeding population should be mated with unrelated superior animals of the same breed.

Out-Breeding

Breeding of unrelated animals either of the same breeds but not having common ancestors for 4-6 generations (out-crossing) or of different breeds (cross-breeding) or even different species (inter-specific hybridisation) is called out-breeding.

This can be further classified as follows:

Out-Crossing

It is the practice of mating of animals within the same breed but having no common ancestors on either side of their pedigree upto 4-6 generations. The offspring (result of mating) is known as an outcross. It is known to be the best breeding method for animals that are below average in milk production and growth rate of beef in cattle, etc. A single outcross may help to overcome inbreeding depression.

Cross-Breeding

It refers to the mating of superior males of one breed with the superior female of another breed. This is done to combine the desirable qualities of the two breeds into one individual.

The hybrid progeny may be used for the commercial production or they may be subjected to some form of inbreeding and selection. This is to develop new stable forms that may be superior to the existing breeds.

A new sheep breed, Hisardale is developed in Punjab by crossing Bikaneri Ewes and Marino Rams.

2.5.3 Interspecific Hybridisation

It is the practice in which animals of one species with the animals of another species, are crossed e.g., mule. A mule is the offspring of a male donkey and a female horse.

In these cases, like mule, the progeny may be of considerable economic value and have combine desirable features of both the parents.

Artificial Insemination (AI)

It is a method of controlled breeding in which the semen collected from a superior male parent is injected into the reproductive tract of the selected female parent by the breeder. The success rate of artificial insemination is fairly low.

The advantages of artificial insemination are:

1. Semen can be used immediately or can be stored and frozen for later use.

2. Semen from a desired breed can be easily transported in a frozen form to distant places where the selected females are present and can be used for impregnating the females on a large scale.

3. It helps in overcoming several problems of normal mating.

2.5.4 Bee-Keeping (Apiculture)

Bee-keeping is also called Apiculture. It includes the maintenance of hives of honeybees for the production of honey and beeswax.

Honey can be used as a food of high nutritive value. And a number of ayurvedic medicines also contain honey.

Beeswax produced by honey bees is used in industry for the manufacture of cosmetics and polishes. Apiculture can be practiced in any area, where there are sufficient wild shrubs, fruit orchards and cultivated crops.

Several species of honeybees are reported in different parts of India, but the most common species reared by bee-keepers is Apis indica.

The important point that should he kept in mind for successful bee-keeping are:

1. Knowledge of the nature and habits of bees.

2. Selection of suitable location for keeping the beehives.

3. Catching and hiving of swarms (group of bees).

4. Management of beehives during different seasons.

5. Handling and collection of honey and beeswax.

Ecological Importance of Bees: Honeybees are the pollinators of many of our crop plants, e.g., sunflower, apple, pear and mustard (Brassica). Hence, keeping beehives in crop areas during flowering period increases pollination and improves honey yield.

2.5.5 Fisheries

A large number of our population is dependent on fish products and other aquatic animals like prawn, crab, lobster, edible oyster, etc. For food, the common freshwater fishes are catla, rohu and common carp.

The marine fishes include, Hilsa, Sardines, Mackerel and Pomfret to meet the increasing demand of fisheries, different techniques like Aquaculture and Pisciculture are applied. Blue revolution is the increased production of fish products. It was being implemented and started in India during the 1960s, along the same lines as 'Green Revolution'.

Economic Importance of Fisheries

Following are the economic importance of fisheries:

(a) It has an important place in Indian economy. It also provides income and employment to millions of fishermen and farmers, especially in coastal states.

(b) A large part of human population depends on fish and fish products.

(c) Fish liver oil is a natural source of vitamin-A. It is used in medicines.

Aquaculture involves the production of useful aquatic plants and animals, such as fishes, prawns, crabs, molluscs (edible and pearl oysters). The practice of fish rearing is called Pisciculture. It involves proper utilisation of freshwater, brackish water and coastal areas.

3

FARM SYSTEMS

3.1 OBJECTIVES

After studying this chapter you will be able to understand:

- Definition of Farming Systems.

- Nature of Farm Level Systems.

- Commercial Agriculture

- Plantation Farming, Dairy Farming and Crop Rotation.

- Farm Types and Structure.

3.2 INTRODUCTION

Farming System is defined as a complex inter related matrix of soil, plants, animals, implements, power, labour capital and other inputs controlled in part by farming families and influenced to varying degrees by political, economic, institutional and social forces that operate at many levels.

Farming system is a decision making tool and land use unit comprising the farm household, cropping and livestock system that transform land, capital and labour in to useful products that can be consumed or sold. Farming system is a resource management strategy to achieve economic and sustained production to meet diverse requirement to farm household while preserving resource base and maintaining a high level of environmental quality. Farming system represents appropriate combination of farm enterprises viz., cropping system, horticulture, livestock fishery, forestry, poultry and the means available to the farmer to raise them profitability. It interacts with environment without dislocating the ecological and socio economic balance on one hand and attempts to meet the food, fibre, fodder and fuel needs as the national goal on other hand.

A farming system is defined as a population of individual farm systems that have broadly similar resource bases, enterprise patterns, household livelihoods and constraints, and for which similar development strategies and interventions would be appropriate. Depending on the scale of the analysis, a farming system can encompass a few dozen or many millions of households (FAO).

'Farming' is a process of harnessing solar energy in the form of economic plant and animal products. 'System' implies a set of interrelated practices and processes organized into functional entity, i.e. an arrangement of components or parts that interact according to some process and transforms inputs into outputs.

Farming system is therefore, designated as a set of agricultural activities organized into functional unit (s) to profitably harness solar energy while preserving land productivity and environmental quality and maintaining desirable level of biological diversity and ecological stability.

The emphasis is more on a system rather than gross output. In other words 'farming system' is a resource management strategy to achieve economic and sustained production to meet diverse requirement of farm household while preserving resource base and maintaining a high level environment quality.

Thus, in farming system all the activities, decision, management, input/output, purchase/sale and resource (s) utilized make the matrix of farming system which interact with socio-economic and bio-physical environment for purchasing the necessary inputs and disposing the outputs by utilizing the natural resources (land, water, air, sunshine etc.) effectively. Sustainability is the objective utilization of inputs without impairing the quality of environment with which it interacts. Therefore, it is clear that farming system is a process in which sustainability of production is the objective.

This overall objective is to evolve technically feasible and economically viable farming system models by integrating cropping with allied complementary enterprises for irrigated, rainfed, hilly and coastal areas with a view to generate income and employment from the farm.

3.3 DEFINITIONS OF FARMING SYSTEMS

An agricultural system is an assemblage of components which are united by some form of interaction and interdependence and which operate within a prescribed boundary to achieve a specified agricultural objective on behalf of the beneficiaries of the system.

This definition is analogous to the general definition of any artificial (i.e., man-made) system of which all managed agricultural systems (including specifically the farm-level systems) form one sub-division.

From a practical production, administration and management point of view, 'all agriculture' can be regarded as consisting of sets of systems at 16 Order Levels or levels of generality. As discussed in Section 1.3. these 16 Order Levels largely constitute a nested hierarchy. This book is concerned with the 12 lowest-order systems, those at farm level, i.e., systems of Order Levels 1 to 12.

There is a large diversity in definitions for farming systems. The most frequently used are these:

- ⊙ A farming system is a decision making unit comprising the farm household, cropping and livestock system that transform land and labor into useful products, which can be consumed or sold.

- ⊙ A farming system is a resource management strategy to achieve economic and sustained production to meet diverse requirement to farm household while presenting resources base and maintaining a high level environmental quality.

- ⊙ A farming system is a set of agro-economic activities that are interrelated and interact with themselves in a particular agrarian setting. It is a mix of farm enterprises to which farm families allocate its resources in order to efficiently utilize the existing enterprises for increasing the productivity and profitability of the farm. These farm enterprises are crop, livestock, aquaculture, agroforestry and agri-horticulture.

- ⊙ A farming system is a mix of farm enterprises such as crop, livestock, aquaculture, agroforestry and fruit crops to which farm family allocates its resources in order to efficiently manage the existing environment for the attainment of the family goal.

- ⊙ A farming system is a unique and reasonable stable arrangement of farming enterprises that a household manages according to well defined practices in response to the physical, biological and socio-economic environment and accordance with the household goals preferences and resources.

- ⊙ A farming system is defined as a complex interrelated matrix of soil, plants, animals, implements, power, labor, capital and other inputs controlled in part by farming families and influenced to varying degrees by political, economic, institutional and social forces that operate at many levels.

- ⊙ A farming system is defined as a population of individual farm systems that have broadly similar resource basis, enterprise patterns, household livelihoods and constraints, and for which similar development strategies and interventions would be appropriate.

There are significant differences among farming systems depending largely on agro-ecological conditions and pedo-climatic zones. This agro-ecological diversity, plus the heterogeneity of economic, political and social conditions has resulted in the development of a wide variety of farming systems.

3.4 GENERAL REQUIREMENTS TOWARDS FARMING SYSTEM CLASSIFICATIONS

For the purposes of iSQAPER we needed a dynamic agricultural production system classification that can be mapped and refined through time.

The farming system should be able to:

⦿ Summarize existing global agricultural production classifications

⦿ Develop a common classification framework that:

a)   Can be mapped using existing global or at least EU/China data sets,

b)   Meet various operational requirements e.g. stratification for data sets, livestock and crop production modelling,

c)   Be of operational level use at EU and China level,

⦿ Develop a detailed plan of work for completing a global system classification;

a)   The definition of the category of mixed farming is challenging,}

b)   To define a generally applicable production system classification,

c)   A particular cropping system may be associated with a number of different livestock system and a particular livestock system may be associated with a number of different cropping systems,

d)   The classification scheme should be interpretable and repeatable, given updates of information, additional data layers and adjustments to classification criteria,

e)   The classification should be dynamic to allow investigation of the likely developments of FS in the future, and how they might evolve in response to global drivers such as population pressure, changes in demand for livestock and crop product and climate change.

3.5 GENERAL SYSTEMS CLASSIFICATION

Discussion and analysis of systems can be of them as actual systems (e.g., of constituent physical processes in the case of natural physical systems) or as representational systems. Common representations or models of actual systems take such forms as written descriptions, physical models, mathematical models, flowcharts, tables of data and computer programs. In the following discussion, reference is to representational systems.

3.5.1 Natural, Social and Artificial Systems

Systems can be classified into three broad families or divisions as either natural, social or artificial systems.

(a)   Natural systems - those that exist in Nature - consist of all the materials (both physical and biological) and interrelated processes occurring to these materials which constitute the world and, inter alia, provide the physical basis for life.

They exist independent of mankind. Our role in relation to natural systems is to try to understand them and, as need be, make use of them. We also (increasingly) attempt to duplicate them, in part or whole; but at this point they become, by definition, man-made or artificial systems. These fundamental natural systems remain unaffected by attempts at imitation. Those natural physical and biological systems which are relevant to agriculture will be self-apparent: rock weathering to form soil; plants sustained by such soil; animals sustained by such plants... are examples of the outward forms of agriculturally relevant natural systems in operation.

(b) Social systems are more difficult to define. Essentially they consist of the entities forming animate populations, the institutions or social mechanisms created by such entities, and the interrelationships among/between individuals, groups, communities, expressed directly or through the medium of institutions. Social systems involve relationships between animate populations (individuals, groups, communities), not between things. Concern here is with human social systems as they relate to or impinge upon farming, and the term social system is used broadly to include institutions and relationships of an economic, social, religious or political nature. There is a certain degree of ambiguity in defining social systems. As an example, the law of property is in its essence a social system. Insofar as it is viewed as consisting of concepts, principles and rules, it is a pure social system, independent of natural systems. But its existence also presupposes the existence of property, including natural physical things, some of which exist as systems. To this extent, as a social system the law of property is dependent on or subordinate to natural systems.

(c) Artificial systems do not exist in Nature. They are of human creation to serve human purposes. All artificial systems, including agricultural systems, are constructed from either or both of two kinds of elements: (a) elements taken from either or both of the other two higher-level orders of systems at division level, i.e., from natural and social systems, and (b) from elements which are constructed or proposed for specific use by each respective artificial system as the need for this arises.

The upper part depicts the dependence relationship between natural and social systems on the one hand and between these and artificial systems on the other. The relevant relationships are: (i) natural systems are independent of systems of the other divisions; (ii) social systems could also be viewed as being independent, but generally a more legitimate view would be that they depend immediately or eventually on natural systems for the essentials of their material existence; and (iii) artificial systems are directly dependent on either or both natural and social systems, or indirectly on natural systems (through the dependence of social systems themselves on natural systems).

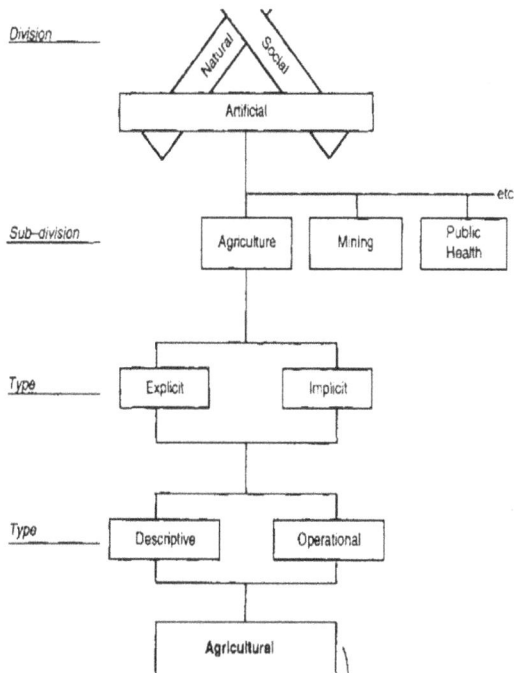

Fig. 1. Agriculture in relation to Other Systems

Agriculture is shown as comprising one of a very large number of actual or potential artificial systems at the sub-division level. Others are those relating to mining, transport, public health, education etc. What such systems at this sub-divisional level have in common is that each is artificial: each is based upon or draws elements from higher-level natural and social systems; and each also contains elements which are purposefully created by some human agency in order to meet its needs.

3.5.2 Further Sub-Classification of Systems

Systems within the three broad divisions or their multitudinous subdivisions can be further classified according to system 'type', a loose term but one which might be used to differentiate among agricultural systems according to a number of factors of which only two are shown in the sketch. As outlined below, first, the system might be either an explicit or implicit one; second, its purpose might be either descriptive or operational. Other 'type' designations could be added; e.g., operational systems could be further classified according to whether or not they are amenable to optimization.

⊙ Explicit systems are those in which the constituent elements are more or less closely identified and defined, and the relationships among these elements are stated formally in quantitative, usually mathematical, terms. Agricultural scientists and economists who work with farmers are concerned mainly with explicit systems. But farmers themselves will seldom be concerned with explicit systems - only with systems of a simpler kind, or only with selected parts of such systems.

- ◉ Implicit systems are systems in which only the main or critical elements are acknowledged and only the major or immediately relevant interrelationships are considered. However, these elements and relationships are not formally recorded, analysed or evaluated. Farmers themselves deal primarily with implicit systems. In more 'advanced' societies, farmers might formalize and work with a few explicit systems or parts of systems (farm record books, simple crop budgets, household expenditure accounts) but here also most agro-management systems will exist by implication.

The purpose in here distinguishing between explicit and implicit systems is to discourage the view that, because farmers (especially small traditional farmers) do not deal with explicit formal systems, these farmers are backward, ignorant, unsophisticated and generally inferior as resource managers. If anything, the facts generally point to a contrary conclusion. While bad farmers can be found anywhere, any close study of small traditional farmers and farming villages in the developing world will, with patience, identify implicit systems at agro-technical, enterprise, farm, farm-household and village levels which are far more complex, sophisticated, sustainable and socially efficient than most agricultural systems found in developed countries.

- ◉ Descriptive systems are usually intended to facilitate an understanding of the organization, structure or operation of a productive process. This might be their sole purpose; e.g., a farmer might construct a simple input-output budget table in order to learn the structural configurations of some potential new crop. Depending on the results of this, he or she might then proceed to construct a more detailed budget (an operational system) to find how best to fit this new crop into his or her farm plan. At higher Order Levels an organogram describing the administrative structure of a ministry of agriculture or of an extension service might be constructed or the flowchart of a commodity from farm to consumer might be drawn - these also are descriptive systems.

- ◉ Operational systems are constructed (by an analyst or manager or research worker) as a basis for taking or recommending action aimed at improving the performance of the system. Such systems are often elaborate. However, increased precision is not infrequently achieved at the cost of decreased practical usefulness. Thus farm managers themselves work primarily with simple operational systems, although the actual physical systems which these represent may be very complex.

As outlined by Dillon (1992), it is also sometimes useful to recognize that, like other systems, agricultural systems may be categorized as:

- ◉ Purposeful or non-purposeful depending on whether or not they can select goals and the means by which to achieve them.

- ◉ Static or dynamic depending on whether or not they change over time in response to internal or external influences.

- ◉ Open or closed depending on whether or not they interact with their environment.

- ◉ Abstract or concrete depending on whether or not they are conceptual or physical in nature.

- ◉ Deterministic or stochastic depending on whether or not their behaviour exhibits randomness over time, i.e., their future behaviour is uncertain.

3.5.3 Agricultural Systems Classification and Order Hierarchy

Agricultural and particularly farming systems exhibit great diversity as shown by, e.g., Duckham and Masefield (1970), Grigg (1974), Kostrowicki (1974) and Ruthenberg (1980). They have been classified in various ways as reviewed by Fresco and Westphal (1988) who also present an ecologically-based classification and typology of farm systems.

The hierarchical classification of farm systems presented here is distinctly different. It is specifically oriented (i) to a farm management and farm-household perspective and (ii) to use as a framework for analysis of what are proposed as the six basic types of farms found in Asia (and elsewhere in the developing world).

Elaboration of the lower part of and relates specifically to agricultural systems. These are listed in largely hierarchical order encompassing 16 Order Levels. Alternatively, with a few minor exceptions, the Order Levels 1 to 16 could have been depicted, reflecting their nested character, as a set of concentric circles with Order Level 1 as the innermost and Order Level 16 as the outermost circle.

The sectoral system, 'all agriculture', is specified as being of the highest order rank, i.e., Order Level 16. Any national or regional agricultural sector, however, consists of such subordinate sub-sectors or subsystems as agricultural credit, education, research, production, transport etc. Each of these constitutes and would be analysed, administered and managed as a system of Order Level 15.

Each such (sub)system may then be further disaggregated into commodity-based industry systems of Order Level 14 such as for coconuts, rubber, wheat, coffee, fish etc. If that flow-path relating to production is being followed, this would then lead to villages or other community units where such production occurs (systems of Order Level 13); these would in turn consist of and could be disaggregated into the individual farm-household systems of order Level 12 which comprise such villages.

Further lower Order Level systems relate to the agro-economic structure of individual farms and, in turn, their component crop and livestock enterprises and to the activities and individual agro-technical processes which underlie such enterprises.

Systems of Order Levels 1 to 12 comprise the field of farm management. But systems of Order Level 1 and 2 are also, indeed primarily, the domain of the applied agricultural sciences. A further proviso is that the 'household' components of farm-household systems of Order Level 12 remain as yet not very well understood. This component is primarily the province of workers in such fields as household economics, rural sociology and social anthropology.

While these various farm family-related fields are fairly well established, they have yet to be brought together in a comprehensive and cohesive way at farm-family level to provide verified models of how rural families in the developing world think about, plan and operate the 'farm' component of their farm-household systems.

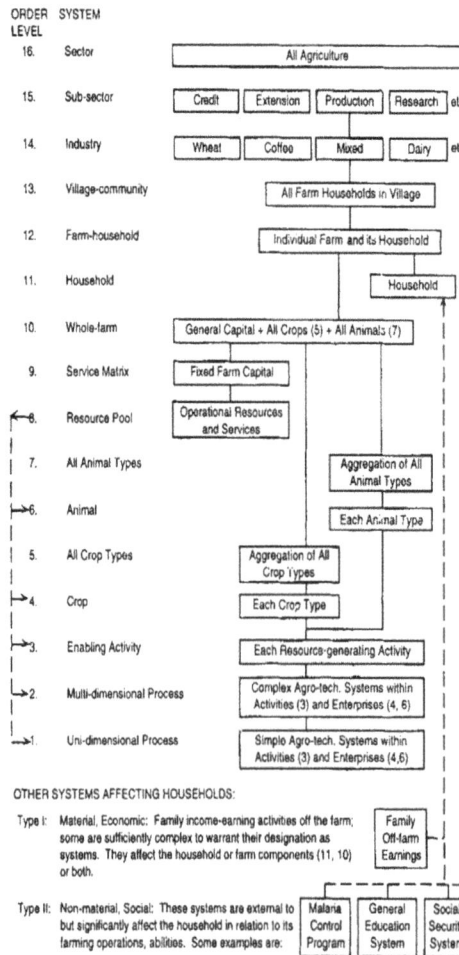

Fig. 2. The Hierarchy of Agricultural Systems

The direction of hierarchical status as proceeding downward from sector to industry to village to farm to crop etc. But whether this direction of subordination is valid will depend on circumstances and analytical purpose. Agricultural scientists would probably reverse the order-ranking shown for the systems on the grounds that, unless the basic agro-technical processes (Order Level 1 and 2 systems) are well developed, the production of individual crops will be inefficient, total farm production will be low and the agricultural sector itself will in consequence be an impoverished one. Similarly extension workers might be inclined to place household systems at the top of the systems hierarchy on the basis that good farming practices (Order Level 1 and 2 systems) will not be adopted unless the household systems are working well, nor consequently will the 'higher'-order systems at industry and sector level operate at their full potential.

3.6 NATURE OF FARM-LEVEL SYSTEMS

The nature of each farm-level system (i.e., Order Levels 1 to 12) of the hierarchy presented may be specified from a management point of view as follows:

- **Order Level 1:** Uni-dimensional process systems. Systems of this lowest order are of an agro-technical nature. They involve an issue or problem which for purposes of analysis or management is abstracted from the context in which it naturally or normally occurs. One example is the application of a single fertilizer element, say nitrogen (N), to a crop and consequent plant response to N in terms of crop yield Y. As noted previously, systems of this order are primarily the domain of physical scientists, but those systems which have practical relevance for farmers thereby also have an economic dimension and so fall within the scope of farm economics. Such simple single-dimensional systems are later examined as processes and as input-output response relationships.

- **Order Level 2:** Multi-dimensional process systems. Systems of this second order are also concerned with limited agro-technical relationships and again they are primarily the domain of physical scientists. They differ from Order Level 1 systems in that they take - or are defined to take - a wider and more realistic view of a subject or problem. To use the same example of fertilizer response: at Order Level 2 an agro-technical system might involve the response of plant growth or yield Y to not one but to several or a large number of input factors such as nitrogen, phosphorous, irrigation water, crop hygiene, soil tilth etc. These multi-dimensional systems also are later examined as processes and as response relationships. Order Level 2 systems can be viewed as aggregations (often interactive) of constituent Order Level 1 systems.

- **Order Level 3:** Enabling-activity systems. Systems of this order are certain enabling activities which generate an intermediate product intended for use as an input/resource by enterprises which do produce a final product. An example is offered by a legume crop turned under to provide fertility for a following (final product-generating) paddy crop. There will often be alternative ways of obtaining this resource: e.g., stripping leaves off leguminous trees, keeping cattle for their manure, or buying a bag of fertilizer. These are all enabling, resource-generating activities but only some of them, the complex ones, warrant designation as systems. They are intended to supply resources to systems of Order Levels 4 and 6.

◉ **Order Level 4:** Crop systems. Systems of this order relate to the production of individual crops; but if these are primarily intended to produce inputs for other crops or livestock, they are regarded as systems of Order Level 3. On many small farms, crop and livestock enterprises produce both final products and resources.

◉ **Order Level 5:** All crop systems. Systems of this order, known also as cropping systems, refer to the combined system of all the individual crops on a farm. On a farm with a single mono-crop, this Order Level 5 system will obviously be equivalent to an Order Level 3 system; but on small mixed farms there will usually be four, five, six or more different crops (of Order Levels 3 and 4) grown in some degree of combination and as many as 20 or more on the highly diversified forest-garden farms of South Asia.

◉ **Order Level 6:** Animal systems. These systems relate to single-species animal enterprises or activities - e.g., dairy cows, camels, fish, ducks. They are the animal equivalent of Order Level 4 (i.e., individual crop) systems.

◉ **Order Level 7:** All animal systems. These systems are the aggregation of all Order Level 6 (sub)systems on a farm. Known as livestock systems, they are the animal equivalent of Order Level 5 (i.e., all crop) systems.

◉ **Order Level 8:** Resource pool. This subsystem is a conceptual device for farm-system planning in which resources and fixed-capital services required by other subsystems are 'stored' in a 'resource pool' from which they are allocated to the other subsystems (of Order Levels 1, 2, 3, 4 and 6). The resource pool is central to operation of the whole farm-household system.

◉ **Order Level 9:** Farm service matrix. A system of this Order Level consists of all the fixed capital resources of a farm which are pertinent to the operation of the farm as a whole but are not assigned to the exclusive use of any particular enterprise or activity: land, fences, barns, irrigation channels and work oxen are common examples. Some of these capital items are true (sub)systems, having interdependence among their component parts (as in an irrigation storage/delivery/distribution network, a grain drying facility, an integrated network of soil conservation structures etc.). Some are only things (e.g., fences, a plough, a barn). But, in its totality, such capital is managed and manipulated as a system for the purpose of providing general services which, while not specific to them, enable the functioning of lower Order Level systems of the farm.

- ◉ **Order Level 10:** Whole-farm systems. Systems of this Order Level consist of all the lower Order Level (sub)systems which go to make up a farm. They consolidate in a single entity all the farm fixed capital, all the operating capital, all the final-product enterprises, all the activities and all the agro-technical processes which underlie such enterprises and activities. Structuring and managing systems of this Order Level are the main tasks or focus of farm management as carried out, on the one hand, by farmers and as investigated, on the other hand, by farm management economists in their professional capacity of providing advice to farm managers, development agencies and governments.

The terms farm system and farming system are often used interchangeably. Here the practice is to use farm system to refer to the structure of an individual farm, and farming system to refer to broadly similar farm types in specific geographical areas or recommendation domains, e.g., the wet paddy farming system of West Java or the grain-livestock fanning systems of Sind.

- ◉ **Order Level 11:** Household systems. On small farms the household itself is the most dynamic and complex of all farm-level systems, although it is a social system not an agricultural one. It dominates the agricultural systems which comprise the farm component. It has two functions: as household it provides purpose and management to the farm component, and as major system beneficiary it receives and allocates system outputs to itself and other beneficiaries.

- ◉ **Order Level 12:** Farm-household systems. These consist of two components or (sub)systems of Order Levels 10 and 11, i.e., the whole-farm system and its associated household system, respectively. The term is a very useful if not mandatory one when used to refer to the small farms of Asia. It carries an insistence that the technical analysis discussed in following chapters will amount to nothing at all unless it is applied to achieving the real needs and aspirations of the household, might be quite a different thing from evaluating the performance of a farm system according to the subjective or preconceived ideas of agricultural technicians and economists. As the peak farm-level system, the farm-household system may be described in system terms as a goal-setting (i.e., purposeful) open stochastic dynamic system with a major aim of production from agricultural resources. These attributes are sufficient to make it also a complex system. The purposefulness of a farm-household system is ensured by its human and social involvement which enables the system to vary its goals and their means of achievement under a given environment. The openness of the farm-household system is obvious from its physical, economic and social interaction with its environment. The non-deterministic or stochastic nature of the farm-household system is guaranteed both by the free-choice capacity of its human (and, if present, animal) elements and by the stochastic nature of the environment with which it (and all its subsystems) interacts. Necessarily, a farm-household system is also dynamic by virtue of its purposefulness, openness and stochasticity which ensure that the system changes over time. Too, any farm-household system is a mixture of abstract and concrete elements or subsystems. The concrete elements are associated with the physical activities and processes that occur in the system. The abstract elements relate to the managerial and social aspects of the system.

3.6.1 Village-Level Farming Systems

Not infrequently in parts of Asia, as also elsewhere in the developing world, the village may replace the farm-household in whole or part as the focal entity for agricultural production. Systems of Order Level 13, i.e., village or community systems, are thus often relevant to the performance of farming systems.

⊙ **Order Level 13:** Village-community systems. Village-level systems or community systems in some situations replace all or part of individual farm-household systems. Three situations are common. First, some production activity in its entirety, including the operation of whole farms as production units, may be on a formal cooperative or group basis. Second, only part of an activity might be carried on by individual farmers while critical parts of it (such as land preparation, the supply of inputs, harvesting and/or marketing) are the responsibility of a formal farmers' club or cooperative. Third, and most difficult to analyse, is the situation found in many Indonesian villages where informal and temporary groups form to perform certain production tasks in common (such as land preparation, irrigation and/or harvesting) then disband and re-form to do different tasks on different crops, with membership continuously changing as individuals drop in and out of groups according to their interests, needs and mutual obligations. In a village there might be 10, 20 or 30 such 'cooperatives', though none might exist officially. Other examples are offered by the semi-nomadic livestock farmers of West Asia who sometimes operate as individual households and sometimes as members of a collective. In all these situations the boundaries of individual units are often so fluid and obscure that the focus for productive analysis has to be the group or village community. (Nevertheless, much externally sponsored farm-development planning remains locked into the mythology of agricultural individualism; perhaps that is why on the small farms of Asia it has borne so little and often poisonous fruit.)

Farm-level systems of Order Levels 1 to 12 are discussed more fully in the following chapters. Before proceeding, however, it will be useful to examine those constituent structural elements of a farm-household system which are relevant to its organization and management.

3.7 STRUCTURAL ELEMENTS OF THE FARM-HOUSEHOLD SYSTEM

The definition of an agricultural system given in Section 1.1 is a general one and applies broadly to systems of all the Order Levels. When applied specifically to a farm-household system of Order Level 12 it implies the system involves ten structural elements or components:

1. Boundaries

2. Household

3. Operating plan

4. Production-enabling resources: the resource pool

5. Final product-generating enterprises

6. Resource-generating activities

7. Agro-technical processes

8. Whole-farm service matrix

9. Structural (interdependence) coefficients

10. Time dimension.

Those elements marked by an asterisk have been considered above as subsystems of the farm-household system. The ten elements are briefly discussed below and, except for structural coefficients and the time dimension, their interrelationships as components of a farm-household system of Order Level 12 are sketched in the example of where they are denoted E1, E2... E8.

1. Boundaries: This first element, the boundaries of the farm-household system, set it apart from other systems and from the world at large. These boundaries are provided partly by the structural characteristics of the particular type of farm, and partly by the purpose of analysis, i.e., to some extent they are subjective and relate to more than the simple physical boundary of the farm. Boundaries are discussed in Chapter 3.

2. Household: As previously noted, the household plays two roles: first, it provides purpose and management to its associated farm system and, second, it is the major beneficiary of its associated farm system. In its first role it provides purpose, operating objectives and management to the farm component of the farm-household system according to its broad domestic and social goals. Obviously these goals vary widely with culture, tradition and the degree of commercialisation and external influences to which the household is exposed. However, one would probably be not too far wrong in offering a generalization

that the primary economic goal on most small farms is security and the primary non-economic goal is social acceptance (Clayton 1983,). If this is correct, the primary objectives for the farm are, first, production of a low-risk sustainable subsistence for primary system beneficiaries; second, generation of a cash income to meet needs not directly met in the form of food and other farm-produced materials; and third, pursuit of both of these in ways which are not in conflict with local culture and tradition.

3. Operating plan: The above objectives are pursued through preparation and execution of a farm operating plan. The core of this may be taken as selection of the best possible mix of agro-technical processes, activities, enterprises and fixed capital (systems of Order Levels 1, 2, 3, 4, 6 and 8).

4. Resource pool: This element was noted above as a system of Order Level 8 central to the management of other subsystems within the farm system.

5. Final product-generating enterprises: These were noted as systems of Order Levels 5 and 7 in the previous section.

6. Resource-generating activities: These also were previously discussed as systems of Order Level 3. They are intended to supplement or entirely supply the resource pool.

7. Agro-technical processes: These were defined above as systems of Order Levels 1 and 2. Processes may be of a biological or mechanical kind. They are a shorthand designation of all the potentially complex and interrelated physical and biological factors underlying production from crop or livestock species, only some of which may be economically relevant.

8. Whole-farm service matrix: This was discussed previously as a system of Order Level 9.

9. System structural coefficients: These coefficients identify and quantify linkage relationships (a) among the various parts or elements within each subsystem and (b) between subsystems. From the general system definition, an essential property of any system is that there be interrelatedness between its parts. In farm-household systems (and in subordinate subsystems of lesser Order Level, particularly Order Levels 4 and 6) such interrelatedness is specified by these coefficients.

10. Time dimension: Unlike mechanical systems which stamp out buttons or TV sets, agricultural systems rest on biological processes which occur over considerable periods of time - from, e.g., a few days in the case of quick-response agricides to 70 or more years in the case of growth and decline of a coconut palm. Agricultural systems are thus inherently stochastic: being dependent on the passage of time, ex ante, their outcomes are uncertain. Moreover, because agriculture is also a set of

economic activities, the old adage applies: time is money. Other things being equal, a system which yields its product or ties up resources over a short time is better than one which yields its output or occupies resources over a long time. Strictly speaking, time is not a system component; rather it is a dimension in which the system operates. The time dimension in relation to resource use is discussed in Section 3.3.4 and in relation to farm planning in Section 9.1. The latter chapter also considers uncertainty as it occurs in farm planning and decision making. Also important from a time perspective are the sustainability and environmental compatibility of the farm system being used. If, over time, the farm system is not biologically and economically sustainable or causes resource degradation, as discussed in Sections 6.2.7 and 8, this is to the disadvantage of both the farm household and society at large.

3.8 STRUCTURAL MODEL OF A FARM-HOUSEHOLD SYSTEM

Before examining the elements of a farm-household system in more detail in later chapters it is useful to consider where they lie in relation to each other in the structure of a small mixed farm.

Element 1, system boundaries: Depending on the purpose of analysis, the farm-household system may be specified with different boundaries. In Figure 3, these are suggested by the shaded circle around the system which sets it apart from other neighbouring systems and from the larger community or environment in which it is imbedded.

Element 2, household: As noted at the top of Figure 3, the household provides objectives and management of the farm-household system and, at the bottom, it exists as the primary internal beneficiary of the system, while distributing some of the system output to external beneficiaries.

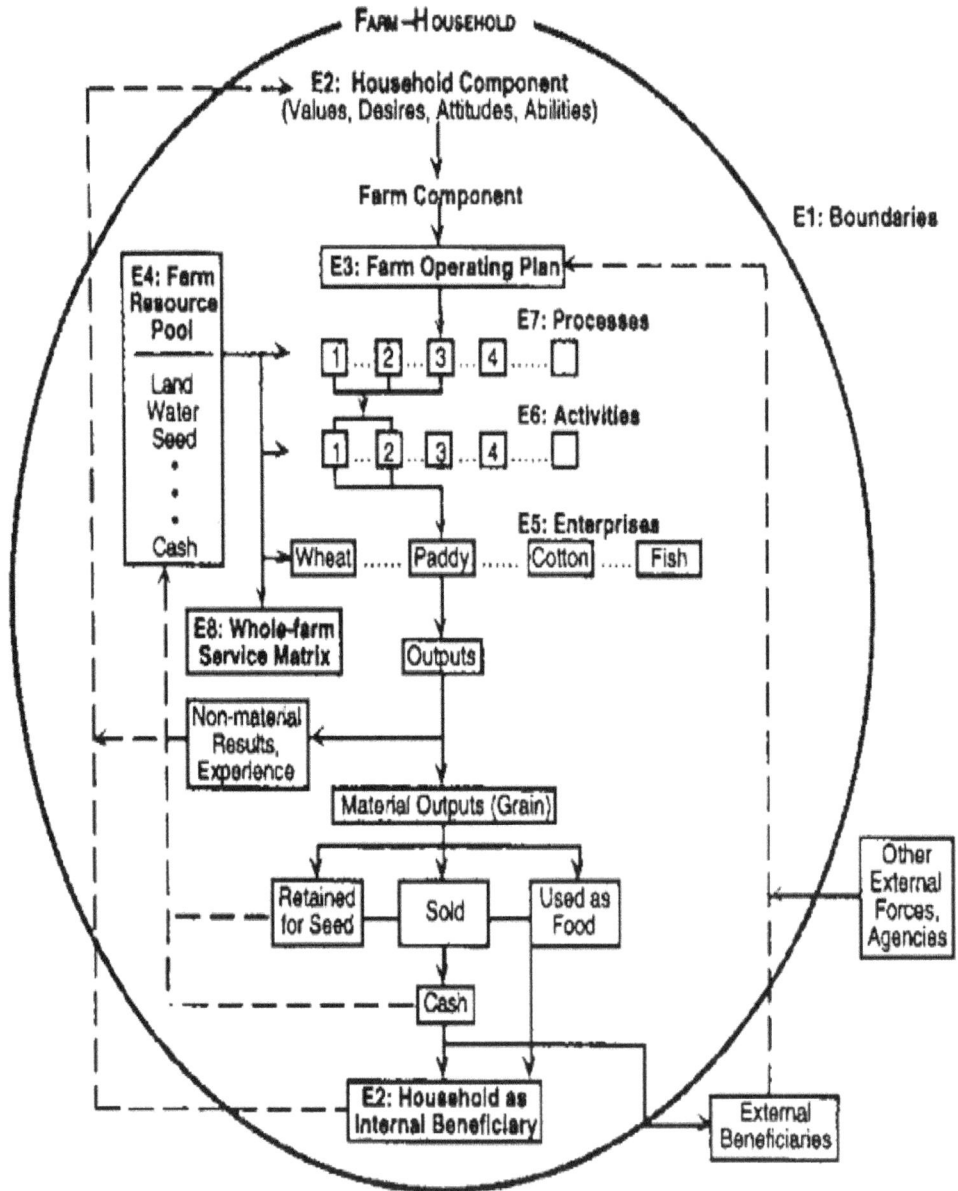

Fig. 3. Interrelationships of Elements in a Simple Farm-household System

Element 3, operating plan: As shown in Figure 3, this is determined largely by the household but it might also be influenced by the requirements of external individuals, agencies or other influences, some of whom might be (external) beneficiaries of the system as outlined below.

Element 4, resource pool: This element consists of resources which are initially present at the time of planning or commencing operation of the system - some pool or stock of land, water, seed, cash etc. which the other elements of the farm system may draw upon. Once the system begins operating, certain components of it (the resource-generating activities and by-products of the enterprises) will replenish the pool. In the schematic sketch of Figure 3, the arrows from the farm resource pool indicate that items from the resource pool flow to processes as well as to activities and to enterprises (as well as possibly to maintenance of the whole-farm service matrix). Strictly speaking, resources should be shown as flowing directly only to processes, since this is the level (subsystems of Order Levels 1 and 2) at which they are actually used. But from a practical viewpoint and because most of the potential processes are actually ignored in planning the operation of a farm system, resources may also be viewed as flowing directly to activities and enterprises as indicated.

Element 5, final-product enterprises: Only four enterprises are shown in the system of Figure 3: wheat, paddy (rice), cotton and fish. Only the flow lines of inputs to and output from the paddy crop are shown.

Element 6, resource-generating activities: In the example of Figure 3, there are two of these supplying some resources to paddy (e.g., a prior fertility-generating legume crop and possibly a cattle activity providing oxpower).

Element 7, agro-technical processes: These underlie the activities and enterprises. Only three are indicated as relevant to paddy in Figure 3 but there are a very large number of biological and mechanical processes actually present in any form of agricultural production.

Element 8, whole-farm service matrix: This was defined previously in relation to a system of Order Level 9. It consists of fixed farm capital which provides a flow of services to all other elements of the system, particularly to Elements 5, 6 and 7 but it is not specific to any one of them.

Element 9, structural coefficients: These are not depicted in Figure 3.

Element 10, time: In the schematic example of Figure 3, the time dimension is not specified explicitly but the model would probably refer to a single operating phase with a duration equal to the life of the longest-term enterprise subsystem, here cotton with a term of six to seven months, or, if climatic seasons are constraining, to a full seasonal cycle of one year. If the system proves to be a 'good' system in terms of household objectives, it might be reactivated in successive phases and continue indefinitely; if a 'very good' system it might permit further farm intensification and development; if a 'bad' system it might be restructured; if a 'very bad' system it will, in the absence of restructuring, prove to be non-sustainable and eventually collapse.

Figure 3 is intended only as a schematic example to indicate the broad relationships which the elements of a farm-household and its associated whole-farm system bear to each other. Real small-farm systems are much more complex than Figure 3 suggests, particularly in their internal cycling of resources.

3.8.1 Farming Systems in India

Farming systems in India are strategically utilized, according to the locations where they are most suitable. The farming systems that significantly contribute to the agriculture of India are subsistence farming, organic farming, industrial farming. Regions throughout India differ in types of farming they use; some are based on horticulture, ley farming, agroforestry, and many more. Due to India's geographical location, certain parts experience different climates, thus affecting each region's agricultural productivity differently. India is very dependent on its monsoon cycle for large crop yields. India's agriculture has an extensive background which goes back to at least 9 thousand years. In India, in the alluvial plains of the Indus River in Pakistan, the old cities of Mohenjo-Daro and Harappa experienced an apparent establishment of an organized farming urban culture. That society, known as the Harappan or Indus civilization, flourished until shortly after 4000 BP; it was much more comprehensive than those of Egypt or Babylonia and appeared earlier than analogous societies in northern China. Currently, the country holds the second position in agricultural production in the world. In 2007, agriculture and other industries made up more than 16% of India's GDP. Despite the steady decline in agriculture's contribution to the country's GDP, agriculture is the biggest industry in the country and plays a key role in the socio-economic growth of the country. India is the second-largest producer of wheat, rice, cotton, sugarcane, silk, groundnuts, and dozens more. It is also the second biggest harvester of vegetables and fruit, representing 8.6% and 10.9% of overall production, respectively. The major fruits produced by India are mangoes, papayas, sapota, and bananas. India also has the biggest number of livestock in the world, holding 281 million. In 2008, the country housed the second largest number of cattle in the world with 175 million.

3.8.2 Climate Effect on Farming Systems

Each region in India has a specific soil and climate that is only suitable for certain types of farming. Many regions on the western side of India experience less than 50 cm of rain annually, so the farming systems are restricted to cultivate crops that can withstand drought conditions and farmers are usually restricted to single cropping. Gujarat, Rajasthan, Punjab, and northern Maharashtra all experience this climate and each region grows such suitable crops like jowar, bajra, and peas. In contrast, the eastern side of India has an average of 100–200 cm of rainfall annually without irrigation, so these regions have the ability to double crop. West Coast, West Bengal, parts of Bihar, U.P. and Assam are all associated with this climate and they grow crops such as rice, sugarcane, jute, and many more.

There are three different types of crops that are cultivated throughout India. Each type is grown in a different season depending on their compatibility with certain weather. Kharif crops are grown at the start of the monsoon until the beginning of the winter, relatively from June to November. Examples of such crops are rice, corn, millet, groundnut, moong, and urad. Rabi crops are winter crops that are sown in October -November months and harvested in February – March. Its typical examples are wheat, boro paddy, jowar, nuts, etc. The third type is Zaid crops which are summer crops. It is sown in February – March and harvested in May – June. Examples are aush paddy, vegetables, and jute.

3.8.3 Irrigation Farming

Irrigation farming is when crops are grown with the help of irrigation systems by supplying water to land through rivers, reservoirs, tanks, and wells. Over the last century, the population of India has tripled. With a growing population and increasing demand for food, the necessity of water for agricultural productivity is crucial. India faces the daunting task of increasing its food production by over 50 percent in the next two decades, and reaching towards the goal of sustainable agriculture requires a crucial role of water. Empirical evidence suggests that the increase in agricultural production in India is mostly due to irrigation; close to three-fifths of India's grain harvest comes from irrigated land. The land area under irrigation expanded from 22.6 million hectares in FY 1950 to 59 million hectares in FY 1990. The main strategy for these irrigation systems focuses on public investments in surface systems, such as large dams, long canals, and other large-scale works that require large amounts of capital. Between 1951 and 1990, nearly 1,350 large- and medium-sized irrigation works were started, and about 850 were completed.

3.8.4 Problems of Irrigation

Because funds and technical expertise were in short supply, many projects moved forward at a slow pace, including The Indira Gandhi Canal project. The central government's transfer of huge amounts of water from Punjab to Haryana and Rajasthan contributed to the civil unrest in Punjab during the 1980s and early 1990s. Problems also have arisen as groundwater supplies used for irrigation face depletion. Drawing water off from one area to irrigate another often leads to increased salinity receiving water through irrigation are poorly managed or inadequately designed; the result often is too much water and water-logged fields incapable of production.

3.8.5 Geography of Irrigation in India

Irrigation farming is very important for crop cultivation in regions of seasonal or low rainfall. Western U.P., Punjab, Haryana, parts of Bihar, Orissa, A.P., Tamil Nadu, Karnataka, and other regions thrive on irrigation and generally practice multiple or double cropping. With irrigation, a large variety of crops can be produced such as rice, sugarcane, wheat and tobacco.

3.9 SHIFTING CULTIVATION

Shifting cultivation is a type of subsistence farming where a plot of land is cultivated for a few years until the crop yield declines due to soil exhaustion and the effects of pests and weeds. Once crop yield has stagnated, the plot of land is deserted and the ground is cleared by slash and burn methods, allowing the land to replenish. Crops like yarn cassava, maize, potatoes are mostly grown This type of cultivation is predominant in the eastern and north-eastern regions on hill slopes and in forest areas such as Assam, Meghalaya, Nagaland, Manipur, Tripura, Mizoram, Arunachal Pradesh, Madhya Pradesh, Orissa, and Andhra Pradesh. Crops such as rain-fed rice, corn, buckwheat, small millets, root crops, and vegetables are grown in this system. Eighty-five percent of the total cultivation in northeast India is by shifting cultivation. Due to the increasing requirement for cultivation of land, the cycle of cultivation followed by leaving land fallow has reduced from 25 to 30 years to 2–3 years. This significant drop in uncultivated land does not give the land enough time to return to its natural condition. Because of this, the resilience of the ecosystem has broken down and the land is increasingly deteriorating.

3.9.1 Shifting Cultivation in Odisha

Odisha accounts for the largest area under shifting cultivation in India. Shifting cultivation is locally known as the podu cultivation. More than 30,000 square km of land (about 1/5 land surface of Odisha) is under such cultivation. Shifting cultivation is prevalent in Kalahandi, Koraput, Phulbani and other southern and western districts. Tribal communities such as Kondha, Kutia Kondha, Dongaria Kondha, Lanjia Sauras, and Paraja are all involved in this practice. Many festivals and other such rituals revolve around the podu fields because the tribals view podu cultivation as more than just a means of their livelihood, they view it as a way of life. In the first year of podu cultivation, tribals sow kandlan (variety of arhar dal). Sowing means spraying the seeds and is used at pre-monsoon time and the area is adequately protected. Yield differs from area to area depending on local climatic factors. After harvest, the land is left fallow. During the pre-monsoon, varieties of rice, corn, and ginger are also sown. Generally, after the third year, the tribals abandon this land and shift to new land. On the abandoned land, natural regeneration starts from the available rootstocks and seed banks. Bamboo comes up naturally; along with many other climbers that regenerate. Generally, this land is not cultivated for the next 10 years.

Frequent shifting from one land to the other has affected the ecology of these regions. The area under natural forest has declined; the fragmentation of habitat, local disappearance of native species and invasion by exotic weeds and other plants are some of the other environmental consequences of shifting agriculture. Areas that have a fallow cycle of 5 to 10 years are more vulnerable to weed invasion compared to 15-year cycles, which have more soil nutrients, a larger variety of species, and higher agronomic yield.

3.10 COMMERCIAL AGRICULTURE

In a commercial based agriculture, crops are raised in large scale plantations or estates and shipped off to other countries for money. These systems are common in sparsely populated areas such as Gujarat, Tamil Nadu, Punjab, Haryana, and Maharashtra. Wheat, cotton, sugarcane, and corn are all examples of crops grown commercially.

3.10.1 Types of Commercial Agriculture

Intensive commercial farming is a system of agriculture in which relatively large amounts of capital or labor are applied to relatively smaller areas of land. It is usually practiced where the population pressure is reducing the size of landholdings. West Bengal practices intensive commercial farming.

Extensive commercial farming is a system of agriculture in which relatively small amounts of capital or labor investment are applied to relatively large areas of land. At times, the land is left fallow to regain its fertility. It is mostly mechanized because of the cost and availability of labor. It usually occurs at the margin of the agricultural system, at a great distance from the market or on poor land of limited potential and is usually practiced in the tarai regions of southern Nepal. Crops grown are sugarcane, rice and wheat.

Plantation agriculture involves a large farm or estate usually in a tropical or sub-tropical country where crops are grown for sale in distant markets rather than local consumption.

Commercial grain farming is a response to farm mechanization and it is the major type of activity in the areas of low rainfall and low density of population where extensive farming is practiced. Crops are prone to the vagaries of weather and droughts and monoculture of wheat is the general practice.

3.10.2 Ley Farming

With increases in both human and animal populations in the Indian arid zone, the demand for grain, fodder, and fuelwood is increasing. Agricultural production in this region is low due to the low and uneven distribution of rainfall (100–400 mm yr"1) and the low availability of essential mineral nutrients. These demands can be met only by increasing production levels of these Aridisols through the adoption of farming technologies that improve physical properties as well as the biological processes of these soils. Alternate farming systems are being sought for higher sustainable crop production at low input levels and to protect the soils from further degradation.

In India's drylands, ley farming is used as a way to restore soil fertility. It involves rotations of grasses and food grains in a specific area. It is now being promoted even more to encourage organic farming, especially in the drylands. Ley farming acts as insurance against crop failures by frequent droughts. Structurally related physical properties and biological processes of soil often change when different cropping systems, tillage, or management practices are used. Soil fertility can be increased and maintained by enhancing natural soil biological processes. Farming provides balanced nutrition for sustainable production through continuous turnover of organic matter in the soil.

3.11 PLANTATION FARMING

This extensive commercial system is characterized by the cultivation of a single cash crop in plantations of estates on a large scale. Because it is a capital centered system, it is important to be technically advanced and have efficient methods of cultivation and tools including fertilizers and irrigation and transport facilities. Examples of this type of farming are the tea plantations in Assam and West Bengal, the coffee plantations in Karnataka, Kerala, and Tamil Nadu, and the rubber plantations in Kerala and Maharashtra.

3.11.1 Forestry

In contrast to a naturally regenerated forest, tree plantations are typically grown as even-aged monocultures, primarily for timber production. These plantations are also likely to contain tree species that would not naturally grow in the area. They may include unconventional types of trees such as hybrids, and genetically modified trees are likely to be used in the future. Plantation owners will grow trees that are best suited to industrial applications such as pine, spruce, and eucalyptus due to their fast growth rate, tolerance of rich or degraded agricultural land, and potential to produce large quantities of raw material for industrial use. Plantations are always young forests in ecological terms; this means that these forests don't contain the type of growth, soil or wildlife that is typical of old-growth natural ecosystems in a forest.

The replacement of natural forests with tree plantations has also caused social problems. In some countries, there is little concern or regard for the rights of the local people when replacing natural forests with plantations. Because these plantations are made solely for the production of one material, there is a much smaller range of services for the local people. India has taken measures to avoid this by limiting the amount of land that can be owned by someone. As a result, smaller plantations are owned by local farmers who then sell the wood to larger companies.

3.11.2 Teak and Bamboo

Teak and bamboo plantations in India are a good alternative crop solution to farmers of central India, where conventional farming is popular. Due to rising input costs of farming, many farmers have grown teak and bamboo plantations because they only require water during the first two years. Bamboo, once planted, provides the farmer with output for 50 years until it flowers. Production of these two trees positively impacts and contributes to the climate change problem in India.

3.12 CROP ROTATION

Crop rotation can be classified as a type of subsistence farming if there is an individual or communal farmer doing the labor and if the yield is solely for their own consumption. It is characterized by different crops being alternately grown on the same land in a specific order to have more effective control of weeds, pests, diseases, and more economical utilization of soil fertility. In India, leguminous crops are grown alternately with wheat, barley, and mustard. An ideal cropping system should use natural resources efficiently, provide stable and high returns, and avoid environmental damage.

3.12.1 Different Sequences of Crop Rotation

Rotation of two crops within a year i.e.:

Year 1: Wheat

Year 2: Barley

Year 3: Wheat again

Three crop rotation i.e.:

Year 1: Wheat

Year 2: Barley

Year 3: Mustard

Year 4: Wheat again

Pearl Millet

Pearl millet crop is mostly grown as a rainfed monsoon crop during kharif (June–July to September–November) and also as an irrigated hot weather (February–June) crop in north, central and south India. Pearl millet is often grown in rotation with sorghum, groundnut, cotton, foxtail millet, finger millet (ragi), castor, and sometimes, in south India, with rice.

On the red and iron-rich soils of Karnataka, pearl millet and ragi rotation are practiced although pearl millet is not always grown annually.

Cluster Bean

Pearl millet crop sequence with crop residue incorporation has significantly increased the productivity in the arid zone of Western Rajasthan where fallow – pearl millet/pearl millet after pearl millet crop sequence is practiced.

In Punjab, the dryland rotation may be a small grain-millet-fallow. In irrigated lands, pearl millet is rotated with chickpea, fodder sorghum, and wheat.

In the dry and light soils of Rajasthan, southern Punjab and Haryana, and northern Gujarat, pearl millet is most often rotated with a pulse-like moth or mungbean, or is followed by fallow, sesame, potato, mustard, moth bean, and guar. Sesame crop may be low-yielding and may be replaced by castor or groundnut.

3.13 DAIRY FARMING

In 2001 India became the world leader in milk production with a production volume of 84 million tons. India has about three times as many dairy animals as the US, which produces around 75 million tons. Dairy farming is generally a type of subsistence farming system in India, especially in Haryana, the major producer of milk in the country. More than 40% of Indian farming households are engaged in milk production because it is a livestock enterprise in which they can engage with relative ease to improve their livelihoods. Regular milk sales allow them to move from subsistence to earning a market-based income. The structure of the livestock industry is globally changing and putting poorer livestock producers in danger because they will be crowded out and left behind. More than 40 million households in India are at least partially dependent on milk production, and developments in the dairy sector will have important repercussions on their livelihoods and on rural poverty levels. Haryana was chosen to assess possible developments in the Indian dairy sector and to broadly identify areas of interventions that favor small-scale dairy producers. A methodology developed by the International Farm Comparison Network (IFCN) examined impacts of change on milk prices, farm management and other market factors that affect the small-scale milk production systems, the whole farm and related household income.

3.13.1 Co-Operative Farming

Co-operative farming refers to the pooling of farming resources such as fertilizers, pesticides, farming equipment such as tractors. However, it generally excludes pooling of land unlike in collective farming where pooling of land is also done. Co-operative farming is a relatively new system in India. Its goal is to bring together all of the land resources of farmers in such an organized and united way so that they will be collected in a position to grow crops on every bit of land to the best of the fertility of the land. This system has become an essential feature of India's Five Year Plans. There is immense scope for co-operative farming in India although the movement is as yet in it infancy. The progress of co-operative financing in India has been very slow. The reasons are fear of unemployment, attachment to the land, lack of proper propaganda renunciation of membership by farmers and the existence of fake societies.

3.13.2 Economics as the Framework for Farm-System Analysis

Economics or economic analysis is the science of making choices so as to best achieve desired objectives given that only limited (physical and other) resources and opportunities are available and that the future is uncertain. There are no choices to which the science of economics cannot be applied. It is just as pertinent, e.g., to the choice of a spouse as to the choice of which crops to grow or to the choice between using an insecticide or using environmentally friendly integrated pest management. In contrast to this wide applicability of economic analysis, financial analysis is restricted to matters that are naturally of a financial or monetary nature. Financial analysis is thus a subset of economic analysis and, in circumstances where everything is valued in money terms, may be the natural way in which to conduct economic analysis. In other cases, it may be feasible to facilitate economic analysis of possible choices by imputing money values to possible gains and losses. And in yet other cases, such as assessing the resource sustainability and environmental compatibility of alternative farm systems, it may often be infeasible to impute money values to the gains and losses of alternative choices. Decisions must then be made using economic analysis based on non-money values, intuition and judgement.

3.14 ALTERNATIVE BASES FOR FARM-SYSTEM ANALYSIS

There are several reasons why farm economics provides a good conceptual framework for most farm-household systems analysis. The most important of these is the necessity to bring the many relationships of a system and between systems to some common unit or basis of comparison. Unless this is done, systems analysis and the comparison of alternatives will not be possible. The base usually most convenient - and in the case of commercial farm systems most relevant and which has the highest degree of universality - is money or financial value. But several other bases for systems analysis are possible and in certain circumstances they might well be more relevant than money value. The four most important bases of comparison are as follows:

(a) **Money Value:** The convenience of using money or financial values as the basis of commercial farm systems analysis will be obvious: it permits the various system inputs (e.g., seed, fertilizer, power, labour etc.) to be standardized as money costs and the various system outputs to be standardized as money returns so that net revenue, i.e., money returns minus money costs, can be used as the basis of comparison between alternatives. In commercial farming, all or most of these inputs/costs and outputs/revenues can be stated in explicit quantitative terms. On the other hand, when dealing with less commercialized systems where there are no actual price-setting markets for farm inputs and/or outputs, one is often obliged to base the analysis on imputed values. However, this is possible only up to a point; beyond this point, as one moves further from a commercial environment towards a traditional one, the attempted use of money value as the basis or numéraire for analysis becomes too abstract to be useful and one has to search for some other base.

(b) **Family Labour Effort:** Probably the best alternative to money value on small family farms of a subsistence or semi-subsistence nature is labour input, both as a measure of inputs and as a yardstick to judge the worth of outputs. At least this is so in the eyes of the majority of Asian (and African) small-farm families for two reasons. First, on these farms most production activities involve few if any commercial inputs and most outputs are also not disposed of through commercial channels. Money hardly enters into the matter at all. What such activities do have as their common factor is family labour - often very hard labour - from hand-preparation of fields, to carrying all inputs/outputs perhaps long distances, to hand-pounding the harvested grain. Not unnaturally then, these families plan, compare and evaluate their several different farming activities and alternatives (i.e., analyse their systems) in terms of labour content. To conduct such analysis on any other basis such as money value would be an incomprehensible abstraction. However, 'labour' is not a simple quantity. It can have several dimensions: quantity when labour is measured in terms of standardized units (e.g., labour-days or task-days on estates); quality where the relevant factor is the actual effort required or the degree of skill or unpleasantness associated with separate tasks; and agency where the labour measurement reflects the social position or status of the person performing the task. Thus, in different societies, patriarchal or matriarchal, women's labour will be valued less or more highly than the labour of men regardless of the actual effort expended, while the labour performed by children might also be valued according to their (usually inferior) social status rather than to the actual work they perform. These dimensions of labour and the implied difficulties of measurement often limit the use of this factor as an alternative to money value. Nevertheless labour often provides a more relevant basis for systems analysis of a very large number of small traditional farms than does money.

(c) **Bio-Mechanical Energy:** A factor which all farm-household systems and their subsystems have in common is their explicit or implicit energy content (including labour, above). Farm-system models have sometimes been structured on the basis of such energy content and inter-component energy flows, Axinn and Axinn (1983). Use of energy-based farm systems analysis rests on the view that, in a world of declining energy resources and materials that can be represented by their energy content, the energy generation and consumption of farm-household systems is a more valid basis for systems analysis than is money profit, and usually also that energy flows which are directly or indirectly involved in all economic activities (including agriculture) are not properly represented - indeed they are often severely distorted - by commercial pricing mechanisms. However, these views involve issues and require solutions at much higher than farm level. Farm systems analysis based on energy flow is more appropriate for some aspects of macro/industry/sector strategic planning than for farm-level operational planning where the immediate interest of farm families is in income (in whatever form it takes) and the effort required to achieve it.

(d) **Water Consumption:** A fourth possible basis for analysis of farm systems is offered by water with systems analysis conducted in terms of the relative water consumption of different crops and animal populations and the implicit water content of products and by-products. Water is obviously the critical common factor in all the farming systems of that great belt of lands stretching from North Africa to India, so much so that even the very wealthy Gulf States, while they have been able to import or create all other agricultural resources, including soils, micro-environments and farmers, remain constrained by water. Moreover, in the 'wet' tropics, the critical nature of this input common to all parts of all farming systems is not yet widely recognized; e.g., even 'well-watered' Java will probably exhaust its water supplies before its soils. However, as important as water is, like bio-mechanical energy it is more appropriate as a basis for some aspects of macro-level systems analysis than for operational-oriented systems analysis at farm level.

In summary, except when used in connection with special-purpose systems, such bases of analysis as energy, water, ecological balance etc. lack the universality and the value orientation required of a general systems base. Money value and labour will probably continue to be used as such a base, either separately in the case of commercial and near-subsistence farms respectively, or jointly in the case of the bulk of small traditional partly commercialized farms.

3.14.1 Farm Management Fields

Farm management analysis and advisory activities can be categorized in terms of four fields defined in terms of the purpose of the analysis as follows:

Field A consists of those problems and analyses which are only or primarily of direct interest to the farmer subjects of the analysis and where solutions to problems are offered on the basis of their beneficial effect on the welfare of these farmers and their families. The great bulk of farm management systems analysis occurs within this field. Field A is the conventional area in which farm management operates, directed to solving the on-farm problems of individuals and groups. Except where otherwise noted, this book is concerned with farm management within Field A; it needs no further discussion at this point except to note that such analysis should, whenever possible, involve farmer participation so as to ensure that the farmer's felt needs are considered.

Field B consists of those problems and analyses which should not really fall within Field A (i.e., they do not properly constitute farm-level problems) but which for convenience or purposes of analysis can be defined, regarded and treated as if they do. Examples of the scope of Field B farm management analysis are offered by the agricultural industries and sectors of some of the mini-states. For example, the agricultural sector of the island nation Kiribati is equivalent to not much more than a single good-sized coconut estate with a few supplementary enterprises added. This sector (a system of Order Level 16) could easily and probably most effectively be analysed, of course with the necessary modifications, as if it were a system of Order Level 10 or 12. An example at industry level is offered by the banana industry of Western Samoa which, although it consists of a large number of individual farms, has been centralized (through the government) in fruit collection/inspection/transport/export and other important aspects. The industry could justifiably be analysed as a single large 'farm system' even though in fact (and in respect of banana-growing activities) it really consists of many farm-household systems.

Obviously, since this type of higher-than-farm-level analysis will be concerned with a range of subject matter in addition to farm economics - processing, marketing, transport, research, extension etc. - farm management can operate successfully in Field B only if the analyst can ignore artificial divisions which are conventionally imposed between the various disciplines. Another condition is that the analysis could not be better performed by a systems analyst working within the conceptual framework of some other discipline.

Field C consists of problems or issues arising within or in relation to higher systems of Order Levels 13 to 16, towards the resolution of which farm management plays only a secondary contributing or partial role, e.g., the lead discipline might be water resource engineering if the problem is planning of a public irrigation project, or agronomy if it consists of planning for the introduction of new crops, or agricultural credit if it is to plan the establishment of a farmers' bank. In this type of supporting role, farm management can operate in any or successively all of Modes 2, 3 and 4, i.e., description, diagnosis and prescription, respectively, as defined in Section 2.1.8 below. However, any 'prescription' that is offered will be of a limited kind and fall short of being a plan for the overall project or program. Analysis will be directed towards the achievement of some global optimum which is not defined in terms of farm management itself. A few of the very large number of situations in which farm management operates in Field C are:

(i) Descriptive studies of farm-household systems to provide background for local or multi/bilateral investment programs in agriculture and/or agricultural infrastructure (e.g., the Country Background Reports of the World Bank). This type of analysis is in Mode 2 (Section 2.1.8 below).

(ii) Diagnostic studies of farm-households to determine just what developmental or investment assistance is needed (e.g., roads, health, transport, extension, credit etc.), and the priority ordering of specific projects to provide such assistance (Mode 3).

(iii) Prescriptive analysis (i.e., Mode 4) aimed at providing the farm-related part of some uni- or multi-purpose project or plan: e.g., farm-level demand schedules for irrigation water in a multi-purpose water storage project (where irrigation is only one planned activity in addition to power generation, flood control, fish production etc.); scheduling of produce supply as part of a feasibility study for the establishment of a cannery.

Field D consists of farm management in the role of generating data for the guidance or support of agricultural policy making. Provision of such data might not require special studies or systems analysis for this particular purpose; often such data will be an incidental output of analysis undertaken for some other purpose, e.g., in Field A. Field D analysis is also of a supportive kind and operates in Modes 2 and 3. The aim is usually to generate knowledge about farm-households or their component subsystems which is to be used by governments, public agencies etc. as a basis for structuring agricultural or broader economic policies - setting farm-input prices or consumer food prices, establishing transport services or credit programs etc. These policies might imply either enhancing farmer welfare or reducing it (e.g., if their thrust is to minimize urban living costs). This is a very important and wide-ranging field: it is difficult to think of any policy which is to affect farmers which should not be based at least in part on farm-level analysis, despite the fact that such farm-level analysis is in fact frequently not carried out, much to the detriment of sound policy making.

3.14.2 Farm Management Modes

Farm management operates in four modes within the above fields.

Mode 1 encompasses routine operational and control activities. It is concerned with the day-to-day operation and management of an actual farm, estate, cooperative or other farm-based producing/marketing entity. This may be thought of as practical or 'muddy-boots' farm management. Management in this mode is largely outside the scope of the present discussion, except that the systems concepts discussed here will, it is hoped, provide principles to guide practical (i.e., Mode 1) management.

Mode 2 refers to descriptive activities whereby farm management provides a conceptual framework for the study, understanding and description of farm systems or farm-related problems. This might be an end in itself; or more likely it will be a necessary stage in the logical-event sequence towards action, as suggested in Figure 2.1. The chief function of descriptive farm management studies is to provide a basis of understanding before problem diagnosis is attempted. There are still many societies in the world with farm-household systems of which we are in nearly complete ignorance; understanding and description of these systems must precede problem diagnosis and, if need be, prescription of solutions.

Mode 3 refers to diagnostic activities concerned with the identification of problems and weaknesses in farm-level systems of all Order Levels 1 to 10 and those parts of Order Level 11 household systems relating to the farm. Such problem diagnosis includes the identification of potential opportunities. Problem diagnosis is usually carried out as a separate mode, but on some commercial farms it might be built into their routine monitoring and management mechanisms (as also on more sophisticated estates).

3.15 ANALYTICAL SITUATIONS WITHIN MODES

In Modes 3 and 4, three analytical situations will arise, viz.:

(i) **Diagnose and prescribe:** First, the problem might require that a diagnostic analysis be made of a system of Order Level from 1 to 10 (or 1 to 12) leading to the identification of some specific weakness or opportunity to be investigated by research on the farm or experiment station. If the problem falls within the competence of the investigator, the analysis would at this point go into prescriptive Mode 4 to develop and offer solutions.

```
Mode 1:      Operation and Control ←──┐
                                      │
Mode 2:      Study                    │
               ↓                      │
             Understanding            │
               ↓                      │
             Description              │
               ↓                      │
Mode 3:      Diagnosis                │
               ↓                      │
Mode 4:      Prescription ────────────┘
```

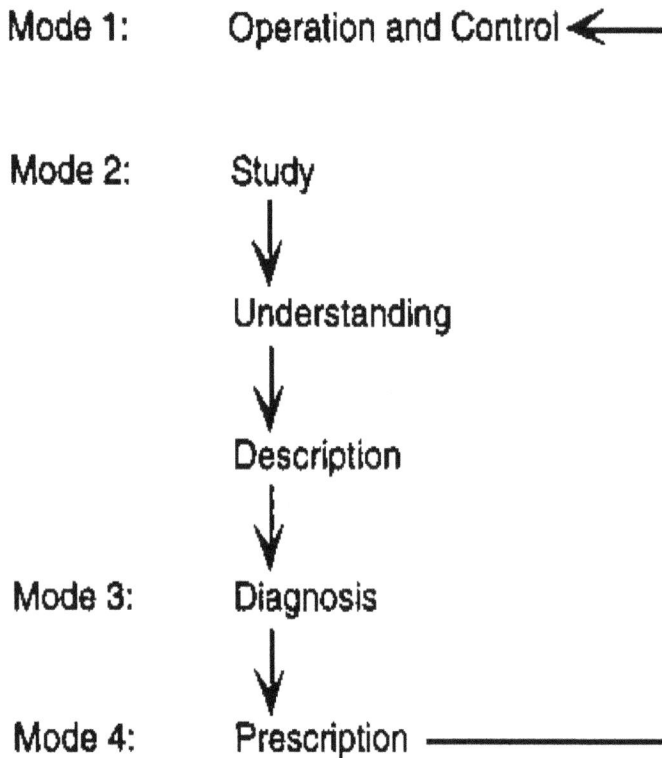

Fig. 4. Relationship between the Four Modes of Farm Management Activity.

(ii) Diagnose and refer: This second situation arises when the diagnosed problem lies beyond the competence of the analyst: e.g., low milk production on a dairy farm might be due to animal disease or a genetic factor or to the household itself (such as a low educational level leading to product adulteration). In such situations the role of the farm analyst is - or should be - to refer the problem to some relevant agency or specialist, i.e., to diagnose but not to prescribe.

(iii) Prescribe only: Often only a prescriptive analysis is called for: e.g., if the problem is to generate land-use plans intended to serve as the agro-economic basis of new settlement projects or transmigration schemes.

3.16 FARM TYPES AND STRUCTURE

Farm structures provide a conducive environment for building a prosperous commercial farm. They are often considered one of the branches of agricultural engineering, dealing with design, elaborate planning, and construction. Thus, you must study all their types and benefits before you begin to grow or breed something.

To date, farms have most often been classified on the basis of agro-ecological factors (such as climate, soil, slope, altitude and, not unrelated to these factors, the crop and livestock systems used) overlaid, to a lesser extent, with socioeconomic criteria. Inevitably such an approach leads to a plethora of farm types. A different approach is taken here. Emphasis is on farm-system structure from a farm management and farm-household perspective with classification based on: (1) the main purpose of the farm, (2) its degree of independence and (3) its 'size'. From such a structural viewpoint there are basically six major types of farm system to be found in Asia and elsewhere around the developing world with dozens of subtypes constituting a continuum of farm types between the extremes of a totally subsistence to a totally commercial orientation.

3.16.1 Farm Types

There are numerous types of agricultural structures used for a variety of purposes. Some of them serve as the home for farmers, their families, and farm staff. Certain farm buildings may be used for keeping crops, livestock or all sorts of equipment. Here are the types of farm structures and their uses.

The six basic farm types are:

Type 1. Small subsistence-oriented family farms.

Type 2. Small semi-subsistence or part-commercial family farms, usually of one half to two hectares, but area is not a good criterion: the same basic structure can be found on much larger 20- to 30-hectare farms as in the Punjab, Sind, and North West Frontier Provinces of Pakistan.

Type 3. Small independent specialized family farms.

Type 4. Small dependent specialized family farms, often with the family as tenants.

Type 5. Large commercial family farms, usually specialized and operated along modified estate lines.

Type 6. Commercial estates, usually mono-crop and with hired management and absentee ownership.

3.17 LAND USE AND LIVESTOCK

Crops and animals are vital elements of the farm structure. The average EU farm has 16 hectares of agricultural land, compared to averages of 180 hectares in the United States, 315 hectares in Canada, and 4 331 hectares in Australia. Altogether, EU farms utilise roughly 157 million hectares of land, of which about one third for growing cereals, slightly less than one third for permanent grassland, and the remaining area for other crops (with industrial crops, permanent crops, and temporary grass and grazing occupying the largest surfaces). Moreover, 5.6 million EU farms with livestock count millions of farm animals – with pigs being the largest group followed by bovines, sheep and goats – plus countless poultry birds as well as other types of animals (e.g. rabbits and horses). On average, they have 21 livestock units (i.e. a reference unit to calculate livestock as the equivalent of one dairy cow). The distribution of land and livestock varies a lot across EU farms, with the smallest farms showing the greatest diversity in terms of on-farm activities.

3.17.1 Need to Dig into the Data

Understanding how farm structure affects the functioning of the farm involves information on such issues as farming specialisations, agricultural practices, agronomic and environmental conditions, and the degree of local development. Therefore, going beyond the main indicators may reveal whether a given farm structure is just right or not adequate at all for a viable farming activity. For example, it may help to explain whether a significant farm workforce is an appropriate labour input or if it stems from a low level of mechanisation or a lack of alternative job opportunities. Also, farms may have large or small acreage, or no land at all, without this accounting, on its own, for strong or weak economic performance. Indeed, farms may have large surfaces because they keep land under cereal production or breed animals on extensive grazing areas. On the other hand, fruit groves or the use of common land (especially for sheep and goat farms) often relate to farms with small land area. The table to the right shows a relevant example of farms with no land area rearing pigs and/or poultry indoors. Although they are not counted as large farms based on hectares of land, these are certainly very large farms based on their high animal numbers compared to the average pig and poultry farm.

3.18 FARM BUILDINGS

These are buildings in the farm, constructed to carry out certain aspects of production functions in the farm business. Farm buildings are beneficial in housing farm workers and farm animals. They also serve as farm offices and stores, and protect workers, livestock, equipment and tools from harsh weather conditions.

Types of Farm Buildings: Some are temporary or make-shift buildings with locally available and cheap materials. Others are stronger, permanent and more costly and complicated buildings.

a. Farm Offices: In large farms, Farm offices serve administrative and management purposes in the farm business for some of the farm staff.

b. Residential Buildings: They provide convenient accommodation for farm staff; for enhanced and quality service delivery.

c. Security Posts: Security personnel and gadgets are housed here. This is to ensure the general security of the farm and farm personnel.

d. Green House: Here, crop plants are raised under controlled environment, especially for experimental, delicate, or out of season plants. The roof is covered with green or glass transparent materials, with regulated temperature.

e. Livestock Pens: These are buildings where farm animals are kept. For example: poultry house, hatcheries for layers, brooder houses for chicks, goat and pig pens, etc.

f. Utility Building: This building serves multi-purpose functions. These include: serving as cold store for storage of fresh fruits and vegetables, meat, fish and poultry products to be prepared for marketing or awaiting consumption. Machine and equipment spare parts, bags of livestock feeds, empty sacs, etc, are also kept there.

3.18.1 Uses of Farm Structures and Buildings

Farm structures and farm buildings are beneficial in the agricultural process in the following ways:

1. They offer proper and suitable environment for processing and storage of farm produce.

2. They provide storage facilities for fertilizers, agro-chemicals, farm tools and equipment.

3. Facilities are made available for housing farm animals and workers temporarily or permanently thus, giving them sense of security.

4. Farm animals and workers are protected from adverse weather conditions.

5. Provision of suitable farm structures and buildings make it convenient for farm workers to care for the farm animals.

6. Farm tools and implements, as well as farm produce, are protected from thieves and predator organisms.

3.18.2 Maintenance of Farm Structures and Buildings

The following activities are recommended for effective maintenance of farm structures and buildings:

i. Destroyed or Damaged fences should be mended or repaired.

ii. Metal storage structures should be coated or painted with aluminum to prevent rusting and reflect light.

iii. Leaking roof should be repaired.

iv. Cracks on walls and floors should be repaired.

v. Broken sewage, water pipes and taps should be mended.

vi. Damaged structures should promptly be replaced.

vii. Floors and walls should be swept clean and disinfected.

viii. Drainage Channels should be cleared regularly to avoid blockage

4

STRATEGIC MANAGEMENT

4.1 OBJECTIVES

After studying this chapter you will be able to understand:

- Strategic Management for Agribusiness.
- Implementation and Control.
- Activities of Farm Management.
- Information Systems for Farm Management

4.2 INTRODUCTION

Strategic management is the integration of all functions so as to pro-actively manage the total farming system in harmony with the internal and external environment towards achieving the long term goals of the farming business. It is about pulling back the lens to get a big picture view and consider future scenarios. It gives you the best opportunity to maintain control, avoid serious pitfalls and capitalize on opportunities. Thinking strategically about your farming business involves creating a vision for where you intend to be in three, five and ten years. Operational Planning is putting the strategy to work through breaking the long- term goals and plans down into shorter terms objectives and action plans. It answers to questions such as "Who is doing what?", "What are the day-to-day activities?", "What are the labour requirements?" and "How do we optimize our resources?".

Farming goals must be developed in direct support of the envisaged future and should focus on key performance areas, such as: Financial goals; Growth/ Expansion goals; Personal goals; Succession planning to ensure a next generation of farmers, and sustainability goals. Each one of these goals must be Specific; Measurable; Achievable; Results-focused and have timelines, hence SMART. Smart goals are easily understood and managed.

4.3 STRATEGIC MANAGEMENT FOR AGRIBUSINESS

Academic research into strategic management dates to the 1950s. The focus of strategic management has shifted from business policy to competitive advantage and finally to corporate governance. Strategic management has also been transformed from focusing on long-range planning, five-force analysis, strategic advantage, core competency, and blue ocean strategy to incorporating flexible corporate strategies appropriate for the rapidly changing modern environment. Earlier studies of agribusiness management have indicated that agricultural researchers have ample opportunities to contribute to the area of strategic management, and these studies have urged meaningful, timely, and applicable research in this critical field.

Strategic management is defined as the process of examining both present and future environments, formulating an organisation's objectives, and making, implementing, and controlling decisions focused on achieving these objectives in present and future environments. Strategic management can also be defined as the process of managing the pursuit of organisational goals while managing the relationship of an organisation to its environment. Dess, Lumpkin, and Taylor (2005) held that strategic management comprises the analysis, decisions, and actions that an organisation executes to create and sustain competitive advantages; in other words, strategic management involves the formulation and implementation of major goals and initiatives by a company's top management on behalf of the company's owners.

Nag, Hambrick, and Chen (2007) indicated that a company's goals and initiatives are formulated and implemented according to the consideration of resources and an assessment of the internal and external environments in which the organisation competes. Academics and corporate managers have developed numerous models to assist in strategic decision making in the context of complex environments and competitive dynamics. These models, including Pearce and Robinson's strategic management model, Porter's forces driving industry competition, the BCG growth-share matrix, and the McKinsey model for business portfolios, have all suggested that both internal and external environments should be considered.

In light of the structural changes in agriculture resulting from climate change and urbanisation trends, the requirements for new entrants, innovation, and social entrepreneurship have become clear. Agricultural practitioners increasingly require entrepreneurship, in addition to sound management and craftsmanship, to be sustainable in the future. On the basis of the aforementioned studies, agricultural attributes, and global sustainability, Rankin et al. (2011) indicated that the strategic management of agribusiness must consider the dimension of sustainability. Therefore, the following literature review is divided into three parts: the external environment, the internal environment, and sustainable development.

4.3.1 External Environment

Traditionally, strategic management models have identified the main elements of external environments as including: the general economy, the regulatory environment, customer markets, competition, suppliers, labourers, and technology. Westgren et al. (1988) identified several external environments that should be considered in agribusiness management, namely the general business environment, industry trends, analysis of competitors, and potential industry entrants. In addition to the domestic market, international agricultural policies critically influence agribusiness performance. Determinants of agricultural export performance may include environmental factors (e.g., hostility and price competition), aspects of export competitive advantage (e.g., firm export competence, export channel knowledge, product adaptation, competitive price, and distributor support), and channel relationship antecedents (e.g., information exchange and cooperation).

Government policy is usually a double-edged sword, creating both advantages and disadvantages for agribusiness. Harling (1994) concluded that government protection is finite, and that business strategies appropriate when a market is protected can be inappropriate when protection is removed. Hartwich and Negro (2010) showed that governments support innovation initiatives through various funding schemes that do not explicitly foster collaboration. Rosairo, Lyne, Martin, and Moore (2012) further indicated that agribusiness management problems may increase if governments do not plan to facilitate their policies.s

The capacity of agribusinesses to respond to changes in industry structure, the competitor landscape, and industry boundaries is essential to maintaining market position. Empirical results suggested that the strategies of product development applied by successful agribusinesses are the key to their faster growth relative to competitors. In addition to research and development (R&D) investment and the risk of failure, agribusiness product-innovation strategies mainly depend on competitors' counterstrategies.

Traditionally, most agricultural firms are family businesses in which family interactions must be incorporated into agribusiness operations. Nowadays, the role of diverse social networks as strategic resources should be noted when considering the external environment of agribusiness management. These networks are created by farmers in various forms for their mutual benefit, such as strategic alliances, joint ventures, partnerships, integration, cooperatives, and value chains. The proper operation of these social networks not only allows farmers to manage and customise their production according to consumer needs but also helps farmers achieve scale economies, reduce transaction costs, accelerate information gathering, share R&D outcomes, expand distribution channels, and eventually promote industrial status.

In summary, the critical external environmental factors for the strategic management of agribusiness are: the general business environment, agroindustrial trends, government policy, international and domestic competition, potential industry entrants, the supply chain, distribution channels, and family interactions.

4.3.2 Internal Environment

The main elements of the internal environment identified by the strategic management literature typically include: business functions (marketing, finance and accounting, production and operation, human resources, R&D, and management information systems), the value chain, and business portfolios. Among the critical issues raised by the agribusiness literature, strategic positioning and goal setting are frequently mentioned. These actions are crucial in helping agribusinesses confront structural change in the industry. Westgren et al. (1988) indicated that agribusinesses with formal planning systems are relatively more concerned with future financial performance, whereas firms with informal planning systems focus more on current measures of financial position. Westgren et al. indicated that product-oriented firms are more sensitive to the potential for identifying new markets than are commodity-oriented firms. In addition, Baker and Leidecker (2001) suggested that the use of a mission statement, long-term goals, and ongoing evaluation are heavily emphasised by high-performing agribusiness.

Providing a quality product can be considered the most essential service of an agribusiness. Russo, Cardillo, and Perito (2003) emphasised the critical roles of R&D investment and the risk of failure in product-innovation strategies. Capitanio, Coppola, and Pascucci (2010) suggested that the capacity to build relationships in product markets is the determinant of product innovation, and that the territorial context determines the relevance of each driving factor of innovation. Liu, Kemp, Jongsma, Huang, Dons, and Omta (2014) further argued that integrative capabilities play a crucial role in innovation novelty, which enhances product superiority, and in improving functional capabilities and gaining market potential. In addition, Gellynck, Cárdenas, Pieniak, and Verbeke (2015) confirmed that trust and innovative entrepreneurial orientation influence farmers' absorptive capacity, and that innovative entrepreneurial orientation influences agribusiness performance.

Marketing is another critical aspect of agribusiness management. A previous study showed that market-oriented agribusinesses are highly innovative and achieve superior performance. A well-balanced marketing plan emphasises overall superiority, and must be designed for both retailers and consumers (Hsu & Wann, 2004). Vertical coordination mechanisms must be installed in the marketing plan, leading to competition among retail chains and thus ensuring quality and building brand equity. Moreover, prior research has indicated that store atmosphere, customer service, and product quality are the main marketing factors that influence customer satisfaction regarding agrifood retailing. Although many farmers have strong marketing preferences associated with traditional spot markets, scholars and practitioners have urged that Internet strategies should be adopted in a supply-distribution management framework, particularly the use of social media and customer relationship management systems for agrifood retailing.

Furthermore, the performance of other aspects of management is also crucial for agribusiness. For example, Harling (1988) indicated that agribusinesses with high returns on assets have more diversified product lines, are superior at controlling general expenses, and have fewer assets. Baker, Starbird, and Harling (1994) determined that factors critical to successfully managing quality in agro-industry are top management, the role of the quality department, employee relations, employee training, and process management. Martinez and Poole (2004) suggested that a move towards flexibility and adopting an entrepreneurial style are both likely to contribute to improved agribusiness performance. In addition, Henderson et al. (2005) indicated that Internet strategies are more likely to be adopted in larger firms with a global scope.

In summary, the critical internal environmental factors for the strategic management of agribusiness include: goal setting, strategic positioning, quality product, service innovation, marketing management, vertical coordination, customer service, top management, employee relations, cost control, and financial management.

4.4 STRATEGIC PLANNING FOR FARM BUSINESSES

Strategic planning involves the development of long-term strategies to increase the profitability and competitiveness of your farm business. This may involve developing new enterprises for your farm such as organic production, on-farm processing, direct marketing of your products to consumers, or the efficient production of traditional farm commodities.

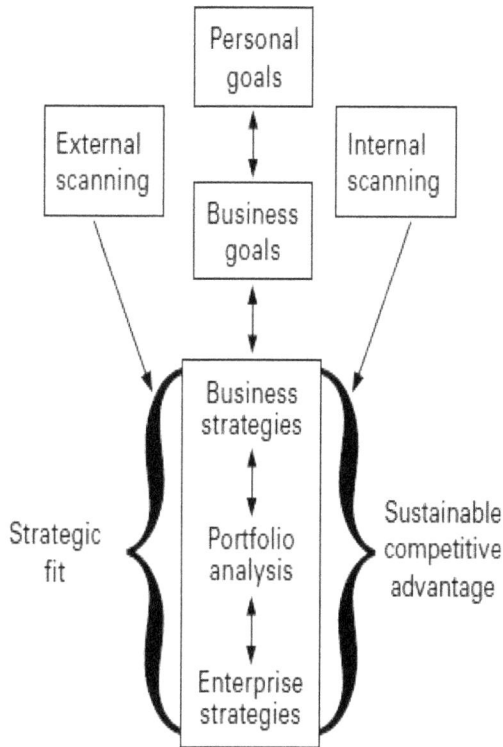

Fig. 1. Strategic planning process.

The purpose of the strategic planning process is to design a farm business that allows the individuals involved in the business to achieve their personal goals. You can do this by using the strengths of your business to take advantage of opportunities in the industry.

Strategic planning involves developing plans for your business and implementing and evaluating these plans. Below is a discussion of a process you can use to develop a strategic plan for your individual farm business. This process is shown in Figure 1.

4.4.1 Phase 1: Factor Analysis

Phase 1 of strategic planning involves assessing and analyzing four factors that are needed for successful strategic planning. These four factors constitute the top half of Figure 1 and provide the ingredients for strategic planning. This first phase of strategic planning requires you to

- ⊙ Identify Personal Goals
- ⊙ Determine Business Goals
- ⊙ Scan the External Environment
- ⊙ Scan the Internal Environment

(1) Personal Goals: One purpose of the business is to achieve the personal goals of the individuals involved in the business. So each person involved in the business should develop personal goals. A personal goal is something that you, as an individual, want to achieve. We often think of making money as our primary personal goal. However, people are complex individuals. Personal goals may include finishing a college degree, spending more time with family, creating a college fund for children, buying a vacation home, getting involved in an organization, doing charitable work, etc. Personal goals may also focus on business activities such as producing safe and nutritious products for consumers, providing employment for the entire family, providing an opportunity for the next generation to farm, etc.

Once your personal goals have been identified, they should be shared with others in the farm business so that inconsistencies among individuals can be identified. Next you need to decide which personal goals will be achieved through the activities of the business. These personal goals will form the basis for developing business goals.

(2) Business Goals: Personal goals provide the foundation for the business goals. This is how the family imposes its wishes on the business. The business goals must be designed to achieve the goals of the individuals and family involved in the business. If business goals are not designed for this purpose, they must be reevaluated.

For example, if the personal goal is to buy a vacation home, the funds may need to come from business profits. So business profitability may be an important business goal. However, if the personal goal is to finish a college degree, the business needs to be structured to provide the time for completing this goal. So the business goal may be to minimize the labor needs of the business or to find an outside labor source.

(3) External Scanning: Scanning is the process of assessing what is going on around you. External scanning involves looking past the farm gate and examining and assessing the economic, business and social environment surrounding your business. It is based on the premise that the environment in which you live and work is not static, but dynamic and constantly changing. External scanning should focus on the following three specific areas.

(a) Industry TrendsL: Identify changes and trends in each of the industries you are competing within. These typically correspond to your business enterprises and may include the corn industry, soybean industry, pork industry, beef industry, etc. However, for value-added farm operations, you may examine market segments or niches of these industries such as organic pork or specialty soybeans.

(b) Competition: Identify and analyze the competition for each of your enterprises. First, divide the competitors into major groupings by size, structure, etc. Next, identify the threat each group poses to your farm business. For example, if you are producing a food product for local consumption, you may be competing against local farmers similar to you. If you are producing for a regional or national market, your competition may include larger food companies.

(c) Economy/Business/Social: Identify changes in the economy, society and the business climate. These would typically include changes such as interest rates, business regulations, inflation, consumer preferences, government programs, demographic distributions, etc. You should examine all changes that will affect your business.

(4) Internal Scanning: Internal scanning involves looking inside of your farm business and identifying its strengths and weaknesses. Business strengths are those things that you do better than your competitors and provides the basis for a competitive advantage. You build a successful business on your strengths. Weaknesses are areas where you are vulnerable to competitors. You look for ways to minimize the impact of weaknesses on your farm business.

Usually the most important assets in the farm business are you and the other individuals in the business. The strengths and weaknesses of a farm business often involve the skills and talents of the people involved in the business. For example, if you are good at networking and working with others, you should try to take advantage of this talent. Conversely, if record-keeping is not your thing, you may want to outsource this function so it doesn't become a weakness of your business.

Scanning should also be conducted for each individual enterprise in the farm business. Traditionally these enterprises have been corn, soybeans, hogs, etc. However, new enterprises are entering farm businesses. Agriculture's traditional broad commodity markets are breaking down into segmented markets of precisely defined agricultural products. For example, you may have a grass-fed beef enterprise or an organic dairy enterprise. Your grass-fed beef enterprise may just involve production or it may include marketing the beef product directly to consumers. The organic dairy enterprise may also include on-farm or jointly-owned processing. This collection of enterprises is what comprises your farm business. So you need to identify and describe each enterprise in your farm business.

4.4.2 Phase 2: Strategy Analysis

After the ingredients for strategic planning have been identified and assessed in Phase 1, Phase 2 involves using these ingredients to strategize alternative ways of organizing the farm business to achieve the business goals. Phase 2 involves the bottom half of Figure 1.

A strategy is the means by which the business uses its strengths (a product of internal scanning) to take advantage of environmental opportunities (a product of external scanning) so the goals identified for the business can be achieved. Because business goals are based on the desires of the individuals in the business, achieving business goals provides an opportunity for achieving personal and family goals.

Strategy development involves identifying a strategic fit between what the business and social environment wants (opportunities) and what your business has to offer (strengths). From this strategic fit, you develop a competitive advantage. A competitive advantage is something that your business can do better than its competitors. For the most value, the competitive advantage should be sustainable over the long-term. This process of using internal and external scanning to identify a strategic fit and develop a sustainable competitive advantage is part of the strategy process identified in Figure 1.

A strategy is developed by first identifying two or more alternatives for your business. This may involve different enterprises or various ways of using resources. From these strategic alternatives, the alternative that best achieves your business goals is chosen.

Business Strategies

The first step in strategy development is to answer two basic questions of the business. What is the planning horizon of the business and what direction is the business going in. You need to consider both a planning horizon strategy and a direction strategy. The planning horizon strategy will often affect the type of directional strategy you choose.

(a) **Planning Horizon** - The planning horizon question asks, how long will the firm exist? Will the business exist for five more years or 25 more years? This is often closely tied to the life cycle of the operator. If you are a young person who recently entered farming, your planning horizon may be quite long and you may be considering an aggressive growth strategy by adding a new enterprise. However, if the business will end in the near future because of your retirement, the planning horizon will be quite short and the strategy may focus on terminating the business. But if the next generation will take over the business, a much longer planning horizon is once again available.

(b) **Direction** - The direction question asks, where is the business going? Will it grow or stay the same size as it is? If it grows, how will it grow? Will you add one or more new enterprises or will you expand your existing enterprises? If you add a new enterprise, will you drop an existing enterprise? Developing a direction strategy should be done in combination with portfolio analysis and enterprise strategy development (below).

Remember, growth strategies don't necessarily mean more acres or more head of livestock. Growth strategies can involve more intensive use of your acres by changing your crop and/or livestock enterprises, or it may involve the further processing and/or marketing of your farm products.

After you have answered these questions you can begin designing the Farm Business Strategy (AgDM File C4-46) for your farm. The primary farm business strategies are:

- ⊙ **Growth** - expanding the size of the business.

- ⊙ **Stability** - maintaining the size of the business.

- ⊙ **Retrenchment** - refocusing the business for improved performance.

- ⊙ **Succession** - transferring the business to younger generation.

- ⊙ **Exit** - ending and leaving the business.

If you plan to grow the business you may want to review the Growth Strategy by type of Farm Business (AgDM File C6-47).

Enterprise Strategy Development

Enterprise strategies, often called competitive strategies, identify how each individual enterprise will compete within its respective market and industry. Enterprise strategies, especially for primary enterprises, are imperative to the success of the business because they link the business and its markets. So enterprise strategies form the foundation for a successful business.

Importance of each enterprise - You defined your business enterprises (above) when you defined your business. Each enterprise should be identified as a profit center, cost center or investment center. Also, each enterprise needs to stand on its own as to its contribution to the business's goals.

- ⊙ **Enterprise interactions** - In addition, enterprise interactions should be examined. What synergies exist among enterprises due to shared resources or management skills that make enterprises complementary? Conversely, are there situations where enterprises compete for resources or management skills?

Portfolio Analysis

Portfolio analysis is part of developing a business strategy. A farm business is made up of one or more enterprises (i.e. corn, soybeans dairy, etc.). Portfolio analysis examines the mix of enterprises in the business and asks the question, what is the best combination and relative size of enterprises for the business. The external scanning exercise should provide you with business and market opportunities. For example, is there demand in your community for locally grown vegetables? How about the expanding demand for organic products? Is their need to start a marketing organization to market farm products directly to consumers?

Next, compare these opportunities to the strengths of the business. Do you have good organization skills? Do you prefer working alone or with others? Are your farm resources suited to producing a new crop enterprise? This exercise involves matching your skills and business resources with business opportunities.

The purpose of this exercise is to develop the type of farming operation that meets the goals of the individuals involved in the farm business. For example, the decision by one of the spouses to pursue off-farm employment may greatly affect how you organize the farm business.

Enterprise Strategy Development

Enterprise strategies, often called competitive strategies, identify how each individual enterprise will compete within its respective market and industry. Enterprise strategies, especially for primary enterprises, are imperative to the success of the business because they link the business and its markets. So enterprise strategies form the foundation for a successful business.

- ◉ Importance of each enterprise - You defined your business enterprises (above) when you defined your business. Each enterprise should be identified as a profit center, cost center or investment center. Also, each enterprise needs to stand on its own as to its contribution to the business's goals.

- ◉ Enterprise interactions - In addition, enterprise interactions should be examined. What synergies exist among enterprises due to shared resources or management skills that make enterprises complementary? Conversely, are there situations where enterprises compete for resources or management skills?

Reality Testing

In a sense, the business and enterprise strategies represent what is possible and the business goals represent what is desired. These strategies can be used to test the reality of the business goals. Reality testing involves examining your business goals in light of the environmental opportunities and business strengths. It addresses the question, can the farm business be organized in a way to achieve the business and personal goals?

4.5 IMPLEMENTATION AND CONTROL

Once your strategy has been selected, action plans or business plans need to be developed of how the strategy will be implemented. Also, a system of evaluation and control needs to be developed to monitor the business and the progress of the strategic plan in achieving the business goals.

4.5.1 Steps to Take Control of Your Farm

1. Get clear on your vision.
2. Create an Enterprise Breakdown
3. Get rid of paper records and create a digital on demand record keeping system
4. Plan and budget pasture and expenses.
5. Dynamically manage stocking rate according to conditions

It's important to involve everyone on your team in this process. It's unlikely that you're a one-man management team, but even if you are, it's wise to bring some other great minds and talents into your proactive analysis and planning.

You probably don't enjoy all aspects of running a livestock business. You may be very skilled and passionate about grazing and livestock management but find financial planning a challenge. If so, you might consider outsourcing your bookkeeping and budgeting.

Not only will involving others in your vision hold you accountable in completing these steps, but you'll be better able to focus on areas of your business where you excel, and both areas will be stronger than before.

Step 1: Get Clear on Your Vision

Before undertaking any challenge, it's best to begin with the end in mind. This step is habit #2 of Stephen Covey's "7 Habits of Highly Influential People", a book that continues to be a best seller 25 years after its release because the habits he describes are crucial for success. Before working in your business one more day, define what the end goal is. Whether you call it a Holistic Goal, a Statement of Purpose, a Vision, or a combination of all three; you need one.

Your ultimate goal should be printed out and posted where you can see it every day. This should be a long term, big picture goal outlining your values and the quality of life you would like for yourself, your children or your loved ones. Studies show that you are 42% more likely to achieve your goals and dreams simply by writing them down on a regular basis. Most of us would buy a lottery ticket every month if the odds were 42%. Why not treat your goals the same way?

Once you've established a broad, long term goal, define the critical milestones you would like to accomplish in the next three years.

- ◉ Try to establish at least ten
- ◉ From here, choose the three most critical goals that you realistically could attain in the coming year (Hint: one of those goals should be implementing the following four steps outlined in this guide.)
- ◉ This process should be repeated annually, and you should have, at the very least, monthly accountability meetings where you establish mini-goals that work toward the broader annual goals.

Creating your list of goals is a great start, but only if they're the right goals; this is where step two comes in.

Step 2: Create a Gross Profit Analysis

The former CEO of Starbucks, Howard Schultz, coined the famous term, "Growth is not a strategy". Many farming businesses view achieving certain economies of scale as the ticket to profitability. Starbucks made the same mistake. They found that haphazard growth in the wrong areas affected their profits in a negative way, so they scaled back and focused on the right kind of growth. They were able to recognize this trend early and reverse the damage because they had objective data to back up the performance of their business model. Without this same objective data that enterprise analysis provides, you may be driving yourself further into debt with enterprises that don't pay, or you may be leaving money on the table by not doing more of the things that are working well in your business.

Define your position by conducting an enterprise breakdown. Think of this as a gross profit analysis of each revenue stream. The goal is to objectively identify the strengths and weaknesses of the current enterprises making up your business model, compare enterprises apples to apples, and to weigh the effects of diverting assets to those enterprises. Examples of separate enterprises might be re-stockers, back-grounding, commercial stock, registered stock, forage production, cropping etc.

Identify Sources of Profit and Loss

The first step in this process is to learn how much or how little the current forms of production are contributing to current profits (or losses). Chances are you'll complete this process already knowing whether you're profitable or not, but the why might be a surprise. Often the seemingly most profitable enterprises also carry with them the bulk of the expenses and producers are shocked to learn that sometimes their main source of income is also driving them farther into debt.

Allocate Expenses

Start by allocating all expenses to one of two categories: Overheads or Variable Expenses.

Overheads are those fixed expenses that you incur simply buy owning this business, whether you produce anything or not. Overheads are things like land payments, insurance, your salary, taxes etc. If you're not sure whether it's an overhead (fixed expense) or a variable expense, ask yourself: "would I have to pay this bill if I didn't have a cow/ewe/goat etc. on this farm?" If the answer is yes, it's an overhead.

Variable expenses include vaccine, fuel, feed, equipment and repairs, utilities, interest and inputs such as fertilizer or seed). Allocate the portions of variable costs to the enterprises to which they belong. When in doubt, call an expense variable: in the long run, no expenses are fixed.

Calculate Total Income

Calculating total income is the next step in this analysis. Allocate income to the enterprises from which they originate.

The hardest part about starting an enterprise analysis is deciding how to organise/ (and) format your calculations. Excel is a great tool because it allows you to allocate a page for each enterprise. An example enterprise breakdown for a farm that owns the land, bales their own hay, and runs a commercial herd of cattle would be to create three separate tabs in an Excel spreadsheet, where you list and calculate income and expenses associated with each income stream.

Land Enterprise	Hay Enterprise	Commercial Cattle Enterprise
+ Revenue () a. Lease to Hay Enterprise b. Lease to Cattle Enterprises c. Rates	+ Revenue a. Hay sales to cattle enterprises (at cost) b. Additional hay sales (at market rates)	+ Revenue a. Calf sales b. Cull cow sales
– Expenses a. Fencing/Water/Other Infrastructure (30-year depreciation) b. Taxes c. Land Payments	– Expenses a. Equipment b. Fuel c. Inputs d. Labour	– Expenses a. Health b. Feed purchased from hay enterprise c. Labour d. Replacement costs e. Breeding costs
= Gross Profit a. Total b. Per Hectare (total ÷ Hectares)	= Gross Profit a. Total b. Per Unit (total ÷ tonnes produced) c. Per Hectare (may help in comparing land use between enterprises)	= Gross Profit a. Total b. Per Unit (total ÷ # cows) c. Per Hectare (may help in comparing land use between enterprises)

The above is a very basic outline that may help you get started. The important thing is to break enterprises down into manageable and quantifiable pieces. A summary page where you can see and compare gross profits of each enterprise may also be helpful in making comparisons. From this basic outline, you can then easily create more production scenarios to compare other forms of production to what you're currently doing.

We recommend RanchingForProfit (US) or GrazingforProfit™ (Australia) as a business changing improvement to cover all the above and more.

Change What's not Working

Find the "leaks" in your vessel, stop bailing water, and get out of the boats that aren't staying afloat on their own. It's tempting to try and simply improve upon what you're already doing or try and add things to optimise the current business structure. Sometimes this works, but sometimes it's like putting a band aid on a bullet wound.

Only you can decide the appropriate course of action for yourself and your operation, but those who are brave enough to question every decision they make, and are bold enough to change what doesn't work, will ultimately have more success in the end. Having this objective data is invaluable to building your analytical skills as a manager.

Step 3: Get Rid of Paper Records and Create a Digital Record Keeping Farm Management System

Paper records have a way of losing themselves in filing cabinets, never to be referenced again. But there are many tools that can make everyday chores faster, easier, and more effective. It's important to keep your records searchable and accessible for future reference.

This is key for several reasons. First, you may have legal obligations to retain certain records for several years. You may also need to access past documents for legal reasons. Finally, your past accounts help facilitate and educate your future decision-making process.

Records are essential to diagnosing economic problems and opportunities, without them, conducting a gross profit analysis and developing a budget is very difficult. Accurate financial and grazing records will improve your ability to manage resources and make business decisions by enabling you to fine-tune your budget and improve the accuracy of your gross-profit analysis.

There are many recordkeeping tools, software and templates that can assist this process. QuickBooks is a dedicated tool for financial management and recordkeeping. MaiaGrazing offers MaiaGrazingLITE which is a powerful FREE platform for keeping grazing records and tracking pasture production, grazing history and rainfall data long term. MaiaGrazingPRO, in addition to recordkeeping, enables you to create multiple grazing plans that are accessible from any device with an internet connection. Alternatively, you can find excellent free Excel grazing planning chart downloads, such as this one, available at holisticmanagement.org.

One advantage to electronic records is the ability to be transparent with agencies, landowners and organisations you may be working with: whether it's the neighbour you lease from, an absentee owner, a consultant or a carbon offset provider. Transparency and communication are key to building trust. Don't let a landowner you lease from wonder whether you're being a good steward of the land. Be proactive and show them with your grazing and production records. You can even implement a monitoring program to track ecological shifts over time.

This kind of proactive communication and transparency will advance your reputation as a responsible land manager and give you a leg up over other producers competing for land leases. You will also be able to more quickly notice ecological shifts and adjust your management accordingly. Remember, the biggest profitability factor in your operation as a grass-based livestock producer is your ability to grow grass.

Step 4: Plan and Budget Pasture and Expenses

Budgeting your finances and pasture helps you to maximise resources, minimise waste and maximise profitability, whether the topic is grass or money – after all, they're one and the same! Before embarking on this process, it's important to remember your vision and goals. Planning ensures you won't resort to old habits that maybe haven't served you well in the past.

Remembering your "why" will help you plan your "what". What actions in regard to your grass and money will get you closer to your goals? The very first step in both financial and grazing planning is to plan what will remain. In the financial planning realm, this is profit. In grazing planning, it's residual grass cover (the final section covers how to account for this). These two elements are the main drivers of sustainability in any farming operation.

Stuart Austin is a MaiaGrazingPRO user who manages 'Wilmot', NSW. He has learned to be a forward thinker who is constantly planning and evaluating his grass management strategy. In late 2017, he was faced with a major medical emergency:

"In late 2017 I landed in hospital in a critical condition, leaving my crew on the farm to manage the best they could for eight weeks, during our busiest time of the season, with over 4,000 head of cattle on hand at the time. One of the ways they managed so well through this period was the use of the Grazing Plans I'd laid out in MaiaGrazing,.

We run an intensive rotational grazing system, moving mobs daily, that requires constant forward planning to ensure we optimise our stocking rate to carrying capacity. There would be a lot more stress around this if it weren't for the simplicity of the MaiaGrazing Planning tool. Combined with the forecasting tool, we are able to plan day-to-day, week to week and month to month, which in an operation like this is critical to minimise stress and optimise efficiency. I don't believe we could run this operation as effectively as we do without MaiaGrazing." ~Stuart Austin

Creating a grazing plan helps to maximize the coordination with all facets of the operation as well as the personal schedules of those charged with implementing the grazing schedule. Before starting a grazing plan, there are three things to consider:

1. Important Personal Events: Mark on the calendar all of the things you want to make time for such as weddings, events, vacations etc. If you plan around the things that are important to you, you're more likely to fit them in.

2. Critical Ecological Areas: Plan to rest, or target, pastures at the appropriate times. If you have a noxious weed problem in one area, try to plan for animal impact at a time when that weed is most palatable or vulnerable to hoof impact. Or, if a riparian area is sensitive to trampling during wet seasons, make note so that turning into that pasture during critical periods won't happen.

3. Peak Nutritional Demand: Save your best forage for seasons such as calving, breeding and weaning. It's during these periods that nutrition is important so either plan your grazing rotation, so animals are consuming quality forage, or have easy access to supplementation, during these periods.

At the beginning of each fiscal year, you should also be sitting down to prepare a budget for the production year. Prepare for bad years, hope for good years, but always count your profit first and align your actions to achieve that profit. Profit is not your salary. Profit is what remains when all overheads, including your salary, and variable expenses have been paid. It should be 10% to 30% of your gross income (income before expenses). The biggest mistake made in budgeting is to take last year's budget and roll the numbers over to the current year. By doing that you're setting yourself up for another year like the last. Unless it was a wonderfully profitable last year, that isn't a good thing! The process of making a budget is much more valuable that the actual finished product. For maximum impact, involve all team members in the budget process to not only get their buy in, but they may have innovative ideas to help cut costs.

Step 5: Dynamically Manage Stocking Rate According to Conditions

Managing "dynamically" means to manage in a way that enables you to quickly adjust to changing variables. For many in agriculture, the primary variable of concern is rainfall. That's why it is beneficial to continuously track your stocking rate per acre per unit of rainfall.

As Bart Davidson of MaiaGrazing will say, "This really just means the rolling ratio of dry matter consumed per inch of precipitation that grew it".

Recognising drought conditions early is relatively easy but changing your management early, and by how much, is a little more difficult. The key to making the most of drought conditions is to react quickly and in an informed way, because the sooner you reduce stock numbers, the smaller the change need be. The longer you wait, however, the greater the change needs to be, and then it's likely you'll also be selling in a falling market. Or if you start feeding, it will almost definitely be at a higher price than if bought early.

Another issue to consider is the protection of the soil surface during drought periods. The target residual biomass cover on the soil surface necessary for ecological health does not change because the land is in drought, so don't count on the "take half/leave half" rule to take proper care of your land. If you notice that bare ground starts to increase with fewer than 800 kg/ha dry matter remaining after a graze, then that should be your target residual each year, regardless of total production. In a good year the land may produce 2,500-3000kg/ha of forage per year and you can remove 1,200 kgs sustainably. In a drought year, if total production is only 1,400 kg's. total, you can now only remove 600 kg's before you start to affect soil health and the sustainability of your farm.

This is where tracking your stocking rate per 100mm of rainfall and planning your grazing in advance according to current conditions can help to take the guesswork out of managing stocking rates. Please don't take the numbers above as recommendations for your land, as every ecosystem is different. Your target biomass residual is specific to your land and your production scenario. Establishing these targets takes observation and management. A healthy landscape produces more, and longer, than a degraded landscape. Ultimately, you're a grass farmer first and livestock producer second.

We hope these 5 Steps have provided you with some first action steps to get systems in place that will help you make smart, forward-thinking decisions and finally get on the trajectory of success. These topics only skim the surface of everything involved in running a successful livestock grazing enterprise so watch for future emails containing valuable information to help you up your management.

4.5.3 The Farmer as Manager

The farmer has two main roles, as manager of the farm but also as a cultivator. The cultivation skills of the farmer are more physical, but the management of the farm is about planning, decision-making and the implementation of plans and decisions.

To manage a farm, the farmer must set goals, plan how to achieve these goals, implement the plans, organise the resources - staff, money, time, land - and keep control over the whole process.

The most important aspect of farm management is decision-making. Decisions can range from which crop to plant, when to irrigate, where to buy livestock, when to sell, which market to supply, which technology is the best to use and how to allocate available labour between farming operations.

Activities of Farm Managements

Farm management can include the following activities:

Identify potential problems and opportunities.

Choosing possible solutions.

Implementing a plan.

Control the progress of the plans and solution.

Coordination of activities to meet the planning objectives.

Evaluate the results of the plan.

4.6 INFORMATION SYSTEMS FOR FARM MANAGEMENT

In order for farmers to manage their farms, they need information systems. Where do farmers get information about their farms? Farm information can come from two sources, from outside the farm (external) and from the farm itself (internal).

External sources of farming information include farming websites, farming apps, magazine articles, farmers' days, study groups, consultants and cooperatives.

Internal sources include the farm's own records and financial statements - two more reasons why recordkeeping in farming is so important.

Farm records usually consist of documents such as a physical inventory, a record of money owed and money that needs to be paid, records of receipts and expenditures, labour records, machinery records and physical production records.

4.6.1 Farming Information Important

Farming information is essential to measure the performance and progress of a farm over time. This information creates a basis to compare current farm performance with previous years. Having farm records and information available can help farmers set goals and evaluate the farm's financial position according to these goals.

It can help the farmer to determine where the strong and weak points of his farm are. Having clear and up-to-date financial records can help with the financial management of a farm, up-to-date financial records of a farm can help a farmer to improve tax planning, establish a basis to apply for credit or financing and to monitor the cash flow position of the farm.

It is crucial to have strict control over the finances of the farm in order to know what has been spent and done at any given time during the year.

4.6.2 Production and Operations Planning

How to benefit from techniques and management tools used in general business

Quality management and control

Production contract evaluation

Discusses how to evaluate contracts and includes a checklist of items that a farmer should consider

Decision-making beyond the traditional microeconomic analysis

In addition to the topics just mentioned, the text includes decision making under risk and the development of scenarios to understand the impact of an uncertain future

5

FARM MANAGEMENT DECISION MAKING PROCESS

5.1 OBJECTIVES

After studying this chapter you will be able to understand:

- Production Decisions.

- Decision Making Process.

- Decision Making Frame work.

- Complexity

5.2 INTRODUCTION

As stated previously, management is decision making, or more precisely, it is determining which alternative will most likely allow the decision makers to achieve their goals. But as this page describes, decision making is more complex than that simple description. The purpose of this page is to consider whether a person can enhance their decision making or managerial skills by thoroughly exploring how decisions are made and implemented.

Agricultural production systems are facing new challenges due to a constantly changing global environment that is a source of risk and uncertainty and in which past experience is not sufficient to gauge the odds of a future negative event. Concerning risk, farmers are exposed to production risk mostly due to climate and pest conditions, to market risk that impact input and output prices, and institutional risk through agricultural, environmental, and sanitary regulations. Farmers may also face uncertainty due to rare events affecting, e.g., labor, production capital stock, and extreme climatic conditions, which add difficulties to producing agricultural goods and calls for reevaluating current production practices. To remain competitive, farmers have no choice but to adapt and adjust their daily management practices. In the early 1980s, Petit developed the theory of the "farmer's adaptive behavior" and claimed that farmers have a permanent capacity for adaptation. Adaptation refers to adjustments in agricultural systems in response to actual or expected stimuli through changes in practices, processes, and structures and their effects or impacts on moderating potential modifications and benefiting from new opportunities. Another important concept in the scientific literature on adaptation is the concept of adaptive capacity or capability. This refers to the capacity of the system to resist evolving hazards and stresses, and it is the degree to which the system can adjust its practices, processes, and structures to moderate or offset damages created by a given change in its environment. For authors in the early 1980s such as Petit (1978) and Lev and Campbell (1987), adaptation is seen as the capacity to challenge a set of systematic and permanent disturbances. Moreover, agents integrate long-term considerations when dealing with short-term changes in production. Both claims lead to the notion of a permanent need to keep adaptation capability under uncertainty. Holling (2001) proposed a general framework to represent the dynamics of a socio-ecological system based on both ideas above, in which dynamics are represented as a sequence of "adaptive cycles," each affected by disturbances. Depending on whether the latter are moderate or not, farmers may have to reconfigure the system, but if such redesigning fails, then the production system collapses.

Some of the most common dimensions in adaptation research on individual behavior refer to the timing and the temporal and spatial scopes of adaptation. The first dimension distinguishes proactive vs. reactive adaptation. Proactive adaptation refers to anticipated adjustment, which is the capacity to anticipate a shock (change that can disturb farmers' decision-making processes); it is also called anticipatory or ex ante adaptation. Reactive adaptation is associated with adaptation performed after a shock; it is also called responsive or ex-post adaptation. The temporal scope distinguishes strategic adaptations from tactical adaptations, the former referring to the capacity to adapt in the long term (years), while the latter are mainly instantaneous short-term adjustments. The spatial scope of adaptation opposes localized adaptation vs. widespread adaptation. In a farm production context, localized adaptations are often at the plot scale, while widespread adaptation concerns the entire farm. Temporal and spatial scopes of adaptation are easily considered in farmers' decision-making processes; however, incorporating the timing scope of farmers' adaptive behavior is a growing challenge when designing farming systems.

System modeling and simulation are interesting approaches to designing farming systems which allow limiting the time and cost constraints encountered in other approaches, such as diagnosis, systemic experimentation (Mueller et al. 2002), and prototyping. Modeling adaptation to uncertainty when representing farmers' practices and decision-making processes has been addressed in bio-economic and bio-decision approaches (or management models) and addressed at different temporal and spatial scales.

The aim of this paper was to review the way adaptive behaviors in farming systems has been considered (modeled) in bio-economic and bio-decision approaches. This work reviews several modeling formalisms that have been used in bio-economic and bio-decision approaches, comparing their features and selected relevant applications. We chose to focus on the formalisms rather than the tools as they are the essence of the modeling approach.

Approximately 40 scientific references on this topic were found in the agricultural economics and agronomy literature. This paper reviews the approaches used to model farmers' adaptive behaviors when they encounter uncertainty in specific stages of, or throughout, the decision-making process. There is a vast literature on technology adoption in agriculture, which can be considered a form of adaptation, but which we do not consider here to focus on farmer decisions for a given production technology. After presenting some background on modeling decisions in agricultural economics and agronomy and the methodology used, we present formalisms describing proactive behavior and anticipation decision-making processes and formalisms for representing reactive adaptation decision-making processes. Then we illustrate the use of such formalisms in papers on modeling farmers' decision-making processes in farming systems. Finally, we discuss the need to include adaptation and anticipation to uncertain events in modeling approaches of the decision-making process and discuss adaptive processes in other domains.

The farm manager has to take decisions on several aspects for profitable operation of the farm. Decisions have to be taken regarding production, marketing and administration. These three segments are interlinked and decision making is also interlinked. In this lecture we shall look into the type of decisions that the farmer has to take and the micro-economics principles that could used for decision making.

Production decisions: Basically the farmer has to decide the following:

- ◉ **What to produce** - The farmer has to decide the crops and allied enterprises that he wishes to produce in the farm. It depends on many factors like soil type, water availability, other resources that the farmer can mobilize, agro-climatic factors in the region and above all the needs of the market. Presently the emphasis is on market oriented agriculture. It would be more beneficial to the farmer if he produces the crop and variety preferred by the consumers, since he can sell them at a good price. Based o the above factors a farmer can narrow down the choice of crops and allied enterprises and then identify the optimum enterprise combination that would generate maximum net revenue.

- ◉ **When to produce** - The timing of release of output in the market is important since generally there is a glut in the market during harvest season and it leads to fall in price and eventually the low profit / loss to the farmer. Natural factors also influence the choice of cultivation of crops during a season.

- ◉ **How to produce** - The farmer has to decide the choice of technology for crop production. The farmer could go for organic farming or integrated approach using both organic and inorganic inputs. The choice and level of different inputs, the mode and timing of their application influence the yield.

⊙ **How much to produce** - The area under a particular crop, size of poultry / livestock enterprise etc., influences the quantum of output. The size of enterprises directly influences the expenditure for cultivation and the farmers should be able to meet them from his own or borrowed resources.

In the case of perennial crops like fruit trees, the decisions have a long term impact on productivity and returns.

Managerial Decisions

⊙ Managerial decisions in a farm include human resource management (hiring and supervision of casual and permanent labour), utilization of funds, accounts and record maintenance, financial transactions, accessing information required for farm management etc.

Marketing Decisions

⊙ Marketing decision includes buying of inputs and selling of outputs. Buying decisions address the questions of when to buy? where to buy? how to buy? and how much to buy?. These decisions are important in determining the profitability of the farm business. Similarly the farmer is also confronted with the questions of when, where and how, to sell his produce? Marketing decisions play a crucial role in making the farm business a success or failure.

5.3 DECISION MAKING PROCESS

Decision making is a process; it involves steps. For example:

⊙ Identify and define the problem or opportunity;

⊙ Identify alternative solutions;

⊙ Collect data and information;

⊙ Analyze the alternatives and make a decision;

⊙ Implement the decision;

⊙ Monitor and evaluate the results;

⊙ Accept responsibility

⊙ Where do you find information about your business? What type of production and financial records are needed? This question is revisited throughout the course.

⊙ A long-time employee of a major agribusiness firm recently stated that "despite what we want, in many situations decisions need to be made with incomplete information; managers need to learn how to recognize what is the most important information for a decision and focus on that."

⊙ Where do you find information about the industry? Some information about the industry will be private and some will be public.

⊙ How is the decision making process impacted by whether the information is public or private?

⊙ How does a manager make a decision? We have already discussed the role of goals in decision making. But what about the other steps in decision making? For example,

⊙ How does a manager "evaluate the results?" Is there any relationship between the step of "evaluating the results" and the manager's goals?

⊙ Is there another step after "monitor and evaluate" other than "accept responsibility?" If yes, what is it and what role does it have in the overall decision making process?

⊙ HINT -- it has already been suggested that the functions of management are "planning, deciding, implementing, controlling"; how does controlling relate to "monitor and evaluate?"

⊙ What does it mean to "accept responsibility?" Why is this step important in the decision making process? Does this step relate to risk?

The following list suggests an alternative description for a decision making process. Do these steps align with the steps in your decision making?

A frequent response to this multiple-step decision making process is that "a manager makes decisions without taking the time to consider each item", or "there is not enough time to consider each item in this list". Those observations are accurate but as a person who may have only limited decision making experience, consider slowing your thought process (yes, slow your thought process), consider each step, and only after you have achieved an understanding of each, accelerate your thought process. Without taking the time to understand the process, you could overlook an important step and reach an "incorrect" decision. Make sure you understand your decision making process, then speed up your thinking.

⊙ Briefly describe the current situation.

⊙ Review long-term or overall business/career/professional and personal goals; for example, consider how personal interests and desired level of risk exposure influence goals.

⊙ Assess whether the current situation is "on track" to achieve long-term goals. If yes, there may be no need for a change at this time. If no, this may be an appropriate time to consider alternatives.

⊙ Identify or summarize why the current situation appears to be "not on track" to meet long-term goals; that is, why the current situation will not achieve the goals. How does reviewing the current situation that is considered "not on track" relate to the concept of "effective," as introduced earlier?

◉ Also, recognize that the current situation could be "on track" to achieve the long term goals, but an alternative could achieve those goals in less time or with fewer resources. How does this relate to the concept of "efficient," as introduced earlier?

◉ State reasons why change may be needed at this time; e.g., goals have been changed/revised; conditions or circumstances have changed or are expected to change; an alternative could achieve the goals more efficiently.

◉ Specifying reasons for wanting to review an alternative should/will help clarify the thought process throughout the review process. That is, if thoughts become confused while reviewing alternatives, refer to the reasons why the review is needed.

◉ If change appears necessary: identify alternatives; determine why these are legitimate alternatives; specify what the alternatives are to accomplish, that is, state goals for the alternatives. Consider how the goals for the alternatives align with long-term or overall goals.

◉ Concisely state the decision that needs to be made, for example, "will purchasing this new item of equipment increase the business' profit?"

◉ Specify criteria for deciding whether to implement an alternative, that is, what is the minimum the alternative must be projected to accomplish before it will be implemented

◉ Long-term goals as well as goals for the alternatives will likely consider objectives other than "maximizing profit."

◉ Concisely state the decision criterion, for example, "the business will purchase this new item of equipment if the analysis projects that it will increase annual profit by $2,000."

◉ If none of the alternatives are projected to accomplish their respective minimums, the current situation may be the best possible strategy at this time.

◉ Identify an appropriate method to analyze the alternative (e.g., enterprise analysis, partial budget, whole-firm); briefly state why this analytical method is appropriate for the analysis.

◉ Does the analysis include "bouncing the idea off of" confidants? Do we have a network of peers or colleagues to share and discuss ideas without concern that they may use our idea in a manner that would harm us?

◉ Identify data needed to complete the analysis; consider sources for the data; review how the data will be used in the analysis and decision process.

◉ What assumptions are being made? What is the implication of the assumptions, for example, if the assumptions are changed, would it alter the decision? If yes, the assumptions must be addressing important facts and probably warrant additional effort to gather information to replace or verify the assumption.

- ◉ Conduct the analysis, assess the outcome of the analysis; make a decision about implementing the alternative based on the outcome of the analysis and the goals for the alternative (that is, the decision criterion); develop a brief explanation for the decision.

- ◉ There is no one correct answer for all situations. Even people in what appears to be similar situations are likely to arrive at different decisions due to different goals, different information, different assumptions, and different resources.

- ◉ Whether the decision is to implement an alternative or continue the current practice, develop a plan of action or implementation; identify the steps needed to implement the decision, establish a timeline for each step. establish benchmarks by which to measure whether progress is acceptable; recognize how benchmarks relate to goals for the alternative as well as long-term or overall goals.

- ◉ Implement the plan (assemble necessary resources, dispose of unnecessary resources)

- ◉ Monitor progress (gather data about the performance of the business as the plan is being implemented, compare performance against benchmarks or goals), revise implementation practices; that is, control the business.

Note the numerous decisions that are embedded throughout this decision process. For example, the decision maker needs to decide in the first step which situation needs to be reviewed at this time, and needs to decide in the second step what personal and professional goals to pursue. Each step in the decision making process could be described as a "mini" decision.

- ◉ Where does risk fit into this decision making process? Is assessing risk a component of each step in the decision making process?

- ◉ Is there a difference between analysis of risk, assumption of risk, and management of risk?

- ◉ What is the difference between capacity/ability to assume risk and willingness to assume risk? How does one's capacity to assume risk change over time and how does one's willingness to assume risk change over time? What is the practical implication of these two trends?

- ◉ More on risk in a subsequent section.

5.4 DECISION MAKING FRAMEWORK

Experience has shown that the most important element of the decision process is the ability of people to assess any idea from their own self-interest. The aspects of an idea, industry or any decision need to focus on 'what's in it for me?' After all you are going to have to live with the consequences of any decision you make.

5.4.1 Compatibility

Ideas that are compatible fit easily with existing practices, knowledge and skills. Compatible ideas require less change to the current household or farm system, and have lower costs. Compatible ideas have less risk of failure.

People are more motivated by ideas that minimise mental and social stress.

It is easiest to start an idea that is highly compatible with what is already being done. It is harder to diversify into a completely new idea that bears no relationship to what is being or has been done in the past.

The compatibility of a new idea is made more attractive if economies of scope exist. That is when the new costs of the new idea are less than the cost of producing it on its own. The opposite is true if there are no economies of scope associated with a new enterprise.

5.5 COMPLEXITY

5.5.1 Simple Ideas are Easier to Adopt

The more complex the idea the greater are the changes needed to fit the idea into to an existing system. As complexity increases the risk of failure increases. As complexity increases the need for, and the costs of gaining additional knowledge increases.

There are many types of complexity including:

- Technical
- Financial
- Market information.

Historically many agricultural enterprises represent relatively low complexity and it has been easy for many people to get into a whole range of agricultural enterprises. What is happening is the complexity of agriculture is increasing and this complexity can become a barrier to entry for many.

The factors that can increase financial complexity are:

- Costs of entry on a seriously commercial scale for many enterprises
- Cost of equipment
- Length of the payback period.

5.5.2 Resource Advantages

Resource advantages give one idea an advantage over other ideas or gives some people an advantage over other people.

There are many resource advantages that play important roles in new ideas including:

- Financial

- ⊙ Physical

- ⊙ Legal, human, organisational and information resources.

Resource advantages may be occur at the farm, district or regional level or at the industry level. For example, a resource does not have to be on-farm to provide advantage. Being close to a market, a transport company providing refrigerated transportation direct to market, a highly competent supplier or a government research station providing small-scale research are resources that other localities or people may not have.

5.6 ABILITY TO OBSERVE

Observing an idea increases knowledge and understanding and increases the ability to make good decisions.

An idea that is easily seen and that has been successful in other places is more likely to be success for you. Being able to see and talk to someone else who has already successfully undertaken an idea improves the confidence of people to adopt the idea.

Agricultural products, like many other products have life cycles. Decisions to diversify are improved if people consider the life cycle stage of the product and market. A mature industry that has been developing for many years is highly observable and much information is available.

The more people involved in doing an activity, the more observable it becomes. For example, more product is produced, more information is known about costs, prices, profits and the associated financial and market risks.

5.6.1 Ability to Trial

A better decision can be made if a person trials ideas first.

In an agricultural sense being able to trial an idea means trying it out on a small scale to test it. By doing this a person can see if it will work for them and they start really understanding an idea. A small trial also reduces the financial risk of trying a new idea.

When an idea is trialled on a small scale it would have high average costs of production. As production increases average costs should begin to fall and can reach a minimum at the right scale. If a trial remains a 'trial' then cost disadvantages occur affecting profitability. Suitable scale to operate competitively needs to be considered at the early stages of the diversification activity.

5.6.2 Making a Decision

A decision about an idea is made by considering on balance the issues of compatibility, complexity and relative advantage.

The observation that another person is successful at an idea does not mean you will be successful as the other person may have resource advantages that you cannot obtain or do not know about.

An idea might be supported by strong resource advantage but be rejected on the basis of complexity beyond what you are prepared to do. When moving through the decision-making process a person reaches a point where a go or no go decision must be taken.

Table 1 provides some examples of questions and supplementary questions that require a yes or no answer. If each question delivers a weak or no response then it would be unwise to make a decision to adopt an idea.

Table 1: Questions that help make decisions on ideas based on the decision-making framework.

Focus questions	Supplementary questions
Why adopt the idea?	What is the motive or reason to consider an idea?
What is our current situation?	How well are we doing with what we already have?
	If we are not positioned well why not?
	If we are, why?
How easy is the idea to add to our current activities?	Is it compatible with the existing situation?
	Is it complex or simple?
	Does it significantly change the existing system?
	Can the idea be observed somewhere else?
	Can the idea be trialled on a small scale?
	Can it be scaled up over time?
	Do we have any resource advantages?
How attractive is the potential idea or activity?	What is the expected net present value and internal rate of return?
	What is the payback period?
	Can we achieve the necessary economies of scale?
	Do we have some economies of scope?
	Can a profit still be gained if:
	• input prices rose 10%
	• interest rates rose 1% and 3%
	• wages rose 10%
	• demand (or production) reduced by 50%
	• can volume and price be used to compensate for these variabilitys?
	Are there big players who could dominate?
How sustainable is this idea?	What essential competencies do we bring to this activity?
	What are our resource strengths?
	Are they sustainable?

Focus questions	Supplementary questions
What and where is the market for this product?	Do we have strong market demand for this product or do we have to create it?
How do we get the product to market?	Do we (or can) develop close relationships with other people in the supply chain?
Who are our customer groups?	Do we have a dialogue with our customer groups & know what they want?
	Is this product going to be a commodity product with low prices and low profit?

5.7 THE DECISION-MAKING PROCESS

The decision-making process on a farm (and, in fact, in any facet of life) can be divided into five steps.

5.7.1 Identify the Problem

There might be a problem such as poor yield in your crops, but what is the source of the problem? Are you using poor quality seeds? Do you plant too early so seedlings are damaged by frost? Is your soil too poor?

A problem can also be an opportunity to grow. Start by collecting information. For example, find out how neighbouring farms are performing - do they use compost or fertiliser, when do they plant, how much do they irrigate and so on.

5.7.2 Investigate Alternative Solutions or Actions

After identifying the problems and the real sources of the problems, alternative solutions should be investigated. Can you change your time of planting, add compost or use better seed cultivars? This should be tested to see what the impact will be - earlier planting or enriching the soil with compost before planting could lead to better growth, which could result in an increase in profits.

5.8 CHOOSE THE BEST ALTERNATIVE

Which of the alternatives are more likely to improve farm performance? Compare the advantages and disadvantages of each alternative in terms of cost, how much labour will be needed, timing and risks. (e.g. compost can be expensive to buy and transport but you have cattle so you could make your own, but this can take longer.)

5.8.1 Implement the Decision

When implementing the best alternative, put it into action. The required resources (e.g. labour, irrigation, fertiliser) should be organised, which implies action. The decision maker should accept responsibility for making it happen. (e.g. appoint someone to collect manure and start making compost, select a time to apply compost and plan for enough labour to apply compost at the right time.)

5.8.2 Evaluate the Result

Monitor the progress to see whether the objectives are achieved. Recordkeeping in farming is essential to keep track of changes and improvements. Note down all changes made and any results that were achieved. (e.g. How much compost was applied, what was the difference in yield compared to the plots where you did not apply compost?)

5.9 STRATEGIC DECISIONS FOR THE ENTIRE FARM

Strategic decisions aim to build a long-term plan to achieve farmer production goals depending on available resources and farm structure. For instance, this plan can be represented in a model by a cropping plan that selects the crops grown on the entire farm, their surface area, and their allocation within the farmland. It also offers long-term production organization, such as considering equipment acquisition and crop rotations. In the long term, uncertain events such as market price changes, climate events, and sudden resource restrictions are difficult to predict, and farmers must be reactive and adapt their strategic plans.

Barbier and Bergeron (1999) used the recursive process to address price uncertainty in crop and animal production systems; the selling strategy for the herd and cropping pattern was adapted each year to deal with price uncertainty and policy intervention over 20 years. Similarly, Heidhues (1966) used a recursive approach to study the adaptation of investment and sales decisions to changes in crop prices due to policy measures. Domptail and Nuppenau (2010) adjusted in a recursive process herd size and the purchase of supplemental fodder once a year depending on the available biomass that depended directly on rainfall. In a study of a dairy–beef–sheep farm in Northern Ireland, Wallace and Moss (2002) examined the effect of possible breakdowns due to bovine spongiform encephalopathy on animal sale and machinery investment decisions over a 7-year period with linear programming and a recursive process.

Thus, in the operation research literature, adaptation of a strategic decision is considered a dynamic process that should be modeled via a formalism describing a reactive adaptation processes.

5.10 ADAPTATION FOR THE AGRICULTURAL SEASON AND THE FARM

At the seasonal scale, adaptations can include reviewing and adapting the farm's selling and buying strategy, changing management techniques, reviewing the crop varieties grown to adapt the cropping system, and deciding the best response to changes and new information obtained about the production context at the strategic level, such as climate (Table 1).

DSP was used to describe farmers' anticipation and planning of sequential decision stages to adapt to an embedded risk such as rainfall. In a cattle farm decision-making model, Trebeck and Hardaker (1972) represented adjustment in feed, herd size, and selling strategy in response to rainfall that impacted pasture production according to a discrete distribution with "good," "medium," or "poor" outcomes. After deciding about land allocation, rotation sequence, livestock structure, and feed source, Kingwell et al. (1993) considered that wheat–sheep farmers in Western Australia have two stages of adjustment to rainfall in spring and summer: reorganizing grazing practices and adjusting animal feed rations. In a two-stage model, Jacquet and Pluvinage (1997) adjusted the fodder or grazing of the herd and quantities of products sold in the summer depending on the rainfall observed in the spring; they also considered reviewing crop purposes and the use of crops as grain to satisfy animal feed requirements. Ritten et al. (2010) used a dynamic stochastic programming approach to analyze optimal stocking rates facing climate uncertainty for a stocker operation in Central Wyoming. The focus was on profit maximization decisions on stocking rate based on an extended approach of predator–prey relationship under climate change scenarios. The results suggested that producers can improve financial returns by adapting their stocking decisions with updated expectations on standing forage and precipitation. Burt (1993) used dynamic stochastic programming to derive sequential decisions on feed rations in function of animal weight and accommodate seasonal price variation; he also considered decision on selling animals by reviewing the critical weight at which to sell a batch of animals. In the model developed by Adesina (1991), initial cropping patterns are chosen to maximize farmer profit. After observing low or adequate rainfall, farmers can make adjustment decisions about whether to continue crops planted in the first stage, to plant more crops, or to apply fertilizer. After harvesting, farmers follow risk management strategies to manage crop yields to fulfill household consumption and income objectives. They may purchase grain or sell livestock to obtain more income and cover household needs. To minimize deficits in various nutrients in an African household, Maatman et al. (2002) built a model in which decisions about late sowing and weeding intensity are decided after observing a second rainfall in the cropping season.

Adaptation of the cropping system was also described using flexible plans for crop rotations. Crops were identified to enable farmers to adapt to certain conditions. Multiple mathematical approaches were used to model flexible crop rotations: Detlefsen and Jensen (2007) used a network flow, Castellazzi et al. (2008) regarded a rotation as a Markov chain represented by a stochastic matrix, and Dury (2011) used a weighted constraint satisfaction problem formalism to combine both spatial and temporal aspects of crop allocation.

5.10.1 Adaptation of Daily Activities at the Plot Scale

Daily adaptations concern crop operations that depend on resource availability, rainfall events, and task priority. An operation can be canceled, delayed, replaced by another, or added depending on the farming circumstances.

Flexible plans with optional paths and interchangeable activities are commonly used to describe the proactive behavior farmers employ to manage adaptation at a daily scale. This flexibility strategy was used to model the adaptive management of intercropping in vineyards, grassland-based beef systems, and whole-farm modeling of a dairy, pig, and crop farm. For instance, in a grassland-based beef system, the beef production level that was initially considered in the farm management objectives might be reviewed in case of drought and decided a voluntary underfeeding of the cattle. McKinion et al. (1989) applied optimization techniques to analyze previous runs and hypothesize potentially superior schedules for irrigation decision on cotton crop. Rodriguez et al. (2011) defined plasticity in farm management as the results of flexible and opportunistic management rules operating in a highly variable environment. The model examines all paths and selects the highest ranking path.

Daily adaptations were also represented with timing flexibility to help manage uncontrollable factors. For instance, the cutting operation in the haymaking process is monitored by a time window, and opening predicates such as minimum harvestable yield and a specific physiological stage ensure a balance between harvest quality and quantity. The beginning of grazing activity depends on a time range and activation rules that ensure a certain level of biomass availability. Shaffer and Brodahl (1998) structured planting and pesticide application event time windows as the outermost constraint for this event for corn and wheat. Crespo et al. (2011) used time window to insert some flexibility to the sowing of southern African maize.

5.10.2 Sequential Adaptation of Strategic and Tactical Decisions

Some authors combined strategic and tactical decisions to consider the entire decision-making process and adaptation of farmers (Table 1). DP is a dynamic model that allows this combination of temporal decision scales within the formalism itself: strategic decisions are adapted according to adaptations made to tactical decisions. DP has been used to address strategic investment decisions. Addressing climate uncertainty, Reynaud (2009) used DP to adapt yearly decisions about investment in irrigation equipment and selection of the cropping system to maximize farmers' profit. The DP model considered several tactical irrigation strategies, in which 12 intra-year decision points represented the possible water supply. To maximize annual farm profits in the face of uncertainty in groundwater supply in Texas, Stoecker et al. (1985) used the results of a parametric linear programming approach as input to a backward DP to adapt decisions about investment in irrigation systems. Duffy and Taylor (1993) ran DP over 20 years (with 20 decision stages) to decide which options for farm program participation should be chosen each year to address fluctuations in soybean and maize prices and select soybean and corn areas each season while also maximizing profit.

DP was also used to address tactic decisions about cropping systems. Weather uncertainty may also disturb decisions about specific crop operations, such as fertilization after selecting the cropping system. Hyytiäinen et al. (2011) used DP to define fertilizer application over seven stages in a production season to maximize the value of the land parcel. Bontems and Thomas (2000) considered a farmer facing a sequential decision problem of fertilizer application under three sources of uncertainty: nitrogen leaching, crop yield, and output prices. They used DP to maximize the farmer's profit per acre. Fertilization strategy was also evaluated in Thomas (2003), in which DP was used to evaluate the decision about applying nitrogen under uncertain fertilizer prices to maximize the expected value of the farmer's profit. Uncertainty may also come from specific products used in farm operations, such as herbicides, for which DP helped define the dose to be applied at each application. Facing uncertainty in water availability, Yaron and Dinar (1982) used DP to maximize farm income from cotton production on an Israeli farm during the irrigation season (80 days, divided into eight stages of 10 days each), when soil moisture and irrigation water were uncertain. The results of a linear programming model to maximize profit at one stage served as input for optimization in the multi-period DP model with a backward process. Thus, irrigation strategy and the cotton area irrigated were selected at the beginning of each stage to optimize farm profit over the season. Bryant et al. (1993) used a dynamic programming model to allocate irrigations among competing crops, while allowing for stochastic weather patterns and temporary or permanent abandonment of one crop in dry periods is presented. They considered 15 intra-seasonal irrigation decisions on water allocation between corn and sorghum fields on the southern Texas High Plains. Facing external shocks on weed and pest invasions and uncertain rainfalls, Fafchamps (1993) used DP to consider three intra-year decision points on labor decisions of small farmers in Burkina Faso, West Africa, for labor resource management at planting or replanting, weeding, and harvest time.

Concerning animal production, decisions about herd management and feed rations were the main decisions identified in the literature to optimize farm objectives when herd composition and the quantity of biomass, stocks, and yields changed between stages. Facing uncertain rainfall and, consequently, uncertain grass production, some authors used DP to decide how to manage the herd. Toft and O'Hanlon (1979) predicted the number of cows that needed to be sold every month over an 18-month period. Other authors combined reactive formalisms and static approaches to describe the sequential decision-making process from strategic decisions and adaptations to tactical decisions and adaptations. Strategic adaptations were considered reactive due to the difficulty in anticipating shocks and were represented with a recursive approach, while tactical adaptations made over a season were anticipated and described with static DSP. Mosnier et al. (2009) used DSP to adjust winter feed, cropping patterns, and animal sales each month as a function of anticipated rainfall, beef prices, and agricultural policy and then used a recursive process to study the long-term effects (5 years) of these events on the cropping system and on-farm income. Belhouchette et al. (2004) divided the cropping year into two stages: in the first, a recursive process determined the cropping patterns and area allocated to each crop each year. The second stage used DSP to decide upon the final use of the cereal crop (grain or straw), the types of fodder consumed by the animals, the summer cropping pattern, and the allocation of cropping area according to fall and winter climatic scenarios. Lescot et al. (2011) studied sequential decisions of a vineyard for investing in precision farming and plant protection practices. By considering three stochastic parameters—infection pressure, farm cash balance, and equipment performance—investment in precision farming equipment was decided upon in an initial stage with a recursive process. Once investments were made and stochastic parameters were observed, the DSP defined the plant protection strategy to maximize income.

6

AGRICULTURE PRODUCTION MANAGEMENT

6.1 OBJECTIVES

After studying this chapter you will be able to understand:

- Agricultural Production.

- Profitability of Agriculture.

- Asymmetry of Investment and risk.

- Factors influencing Agricultural Growth.

6.2 INTRODUCTION

Global attention has been devoted to water scarcity and its effect on Indian farmers. However, new analysis from Indian researchers suggests that far more good could come if irrigation were combined with seed improvement.

Tata Trusts and Copenhagen Consensus have commissioned new research by noted experts from India and around the world, looking at measures that would help Indian states respond to major challenges and improve their competitiveness, economic performance, and the well-being and prosperity of citizens. The new research focuses on establishing how much different policies would cost, and what they would achieve overall in economic, environmental and societal benefits.

Now, two new research papers add to the volume of evidence on how to boost agricultural performance. The first of these is by Dinesh Kumar, executive director of the Institute for Resource Analysis and Policy (Irap), Hyderabad. It examines policies that would reduce the effects of water scarcity in Rajasthan and Andhra Pradesh.

In Andhra Pradesh, the Rayalaseema region is hot and dry, with frequent droughts. Only about one-third of the crops are irrigated, and the rest are dependent on rain-fed cultivation, which is susceptible to the vagaries of the weather. Tanks are an important source of water for the rural economy, but—as in other areas—an explosion of well-irrigation has reduced the surface run-off into these tanks. The biggest victims are poor, small, marginal farmers, who depend on tanks for supplementary irrigation for their kharif crop.

There are major water transfer projects being implemented in Andhra Pradesh. This approach—moving surplus water into the tanks, so that they are full—ensures farmers can continue crop production when the tanks do not receive inflows. According to one estimate, the additional storage space available during a drought year is about 1,700 million cubic metres.

The annualized cost of the infrastructure and drainage required to fill the tanks is estimated to be about Rs4,500 per hectare, as well as another Rs2,000 for the annual operation and maintenance of the system. Assuming that the additional water will be used to irrigate around 65,000 hectares of paddy cultivated during winter, the overall annualized cost would be Rs43.2 crore.

Farmers, however, will earn more: The annual incremental net return is estimated to be about Rs9,000 per hectare. There would be further indirect benefits from energy savings because farmers wouldn't need to pump groundwater, as well as from the incremental return from the increase in yield of wells and consequent expansion in the area served, and more intensive watering of irrigated crops. These benefits together add up to Rs15,000 per hectare per year, and the total annual benefits would be Rs159.2 crore.

This means that every rupee spent on the policy in Andhra Pradesh would generate benefits worth nearly four rupees. In Rajasthan, an analysis looking at the same approach but taking into account the specific conditions there, finds there would be benefits worth three rupees.

Kumar also examines state-specific policies: In Rajasthan, renovating the traditional water harvesting system would return three rupees for every rupee invested, while, in Andhra Pradesh, investment in drip irrigation and mulching of high-value crops would generate about five rupees.

These are all respectable returns. But new research by agricultural economist Surabhi Mittal, independent consultant and non-resident fellow, Tata-Cornell Institute for Agriculture & Nutrition (TCI), Technical Assistance and Research for Indian Nutrition & Agriculture (Tarina), suggests another approach may help farmers a lot more.

The researcher looks at various methods of improving farm productivity and farmer income. One of these aims to help solve the problem of the high cost and unavailability of labour through an increase in the level of mechanization by using custom hiring centres, using public-private partnerships. Another approach focuses on relying on information and communication technology (ICT) enabled extension services, which play a crucial role in supporting agricultural activities by taking research, technology and know-how to farmers to improve adoption. Third, the author looks at improving soil health; and, finally, considers improving the availability of certified seeds.

This last idea would generate powerful returns. In India, farm-saved seed from previous crops remains the most prominent source of seeds, year after year, accounting for nearly three-quarters of all seed usage. This means low crop productivity as optimal yield potential is a function of the quality of seeds used.

Although many improved varieties of seeds have been released for cultivation, their full impact has not been realized owing to poor adoption rates as well as poor seed replacement rates.

The solution to this challenge involves spending money on producing more quality seeds (for all the major crops in each state) and promoting these among farmers. The cost over five years adds up to around Rs400 crore in Andhra Pradesh and Rs584 crore in Rajasthan.

But this will lead to better crop yields, increased production, and higher incomes. After reviewing the best evidence, the researcher suggests that yield gains of 10% can reasonably be expected with improved seed replacement rates. Even with this highly conservative assumption, the investment has huge pay-offs: Every rupee spent will have benefits to Andhra Pradesh worth around 15 times the costs, and 20 times in Rajasthan.

Improving agricultural productivity is important in order to improve farmer incomes, and it requires increases in yield, better productivity through the efficient utilization of resources, reduction in crop losses, and ensuring that farmers receive fair prices for output.

The phenomenal benefits from focusing on improving access to seeds highlight the need to prioritize policies that will achieve the most for farmers.

Agricultural products are usually measured by weight or volume. An immediate question arises as to how to best combine different agricultural products since summing over weights or volumes is not very meaningful. One approach when dealing with crops is to convert them to a common physical unit, such as wheat units. More commonly, aggregate output in agriculture is measured in monetary units as the sum of the value of all production in the agricultural sector minus the value of intermediate inputs originating within the agricultural sector. Both cash and non-cash (barter, trade and self-consumption) transactions of final products should be included. This is referred to as "final output" and differs from agricultural GDP by not subtracting out the value of non-agricultural inputs. In other words, final output is the amount of agricultural output available for the rest of the economy, while agricultural GDP measures the net contribution of agriculture to the GDP of a country.

Productivity measures are subdivided into partial or total measures. Partial measures are the amount of output per unit of a particular input. Commonly used partial measures are yield (output per unit of land), labour productivity (output per economically active person (EAP) or per agricultural person-hour). Yield is commonly used to assess the success of new production practices or technology. Labour productivity is often used as a means of comparing the productivity of sectors within or across economies. It is also used as an indicator of rural welfare or living standards since it reflects the ability to acquire income through sale of agricultural goods or agricultural production.

Partial measures of productivity can be misleading, as there is no clear indicator of why they change. For example, land and labour productivity may rise due to increased use of tractors, fertilizer or output mix (move to high value crops). To account for at least some of those problems a total measure of productivity, the Total Factor Productivity (TFP) was devised. TFP is the ratio of an index of agricultural output to an index of agricultural inputs. The index of agricultural output is a value-weighted sum of all agricultural production components. The index of agricultural inputs is the value-weighted sum of conventional agricultural inputs. These generally include land, labour, physical capital, livestock and chemical fertilizers and pesticides. Growth in TFP is referred to as the Solow residual. It is generally considered a measure of technological progress that can be attributed to changes in agricultural research and development (R&D), extension services, human capital development such as education and physical, commercial infrastructure, as well as government policies and environmental degradation. Change in TFP can also be due to unmeasured inputs or imperfectly measured inputs.

6.3 PROFITABILITY OF AGRICULTURE

Estimates for 2016 suggest a muted recovery, specifically in the second half of the year, after sterling weakened relative to the euro. For 2017 so far, prices and markets suggest that the year is likely to be more profitable again, based on current market values, crop and stock conditions.

Total Income From Farming (or TIFF) is a useful measure to look at the UK farming industry as a whole, providing a simple measure of the profit of 'UK agriculture'. In technical terms, TIFF shows the aggregated return to all farmers in UK agriculture and horticulture for their management, labour and their own capital in their businesses.

Examining TIFF in real terms over the long term puts the 2015 decline into perspective as, only seven years earlier, did the industry come to the end of a 10-year period of considerably poorer farming conditions. Whilst difficult, that period made the industry more efficient, stronger and resilient.

The EU's Basic Payment subsidy is calculated in euros then converted into sterling using the average monthly exchange rate for September each year. The 2016 Basic Payment was about 19% better than 2015 because of a more favourable exchange rate. Volumes of output (e.g. milk and cereals) actually fell in 2016 compared with 2015. As it currently stands, profits for 2017 could return to the levels seen in 2011 to 2014, driven by the weaker pound and brisk sales of agricultural commodities since June 2016, where agricultural stocks have been keenly sold to European buyers.

In fact, commodities, by definition, are priced at the point of sale (or close to it if they are sold with a supply contract). Wheat, for example, can vary in price several times throughout a single day. This led to a sharp rise in the value of agricultural outputs since the European referendum in June 2016 (wheat is worth roughly £20 per tonne more and milk about 2.5ppl more).

This rise is an immediate sign of inflation. Farm inputs are mostly products and services. These categories of goods tend to have fixed prices over a longer period of time than commodities: Agrochemical prices are usually adjusted seasonally, wages tend to change annually and other costs such as rents might be reviewed less often than that. In other words, agricultural inputs are generally slower to respond to inflationary pressures than outputs.

This suggests that farm outputs have already benefitted from the decline of sterling, but the costs (assuming sterling remains more or less where it is) will be rising now. We could therefore conclude that, unless agricultural markets fundamentally move up again, the UK farm profitability will fall in 2018.

The nature of commodities and farming means that farmers are largely 'price takers'. Careful cost control can be the key to maximising profits in fluctuating market conditions.

Where incomes and profits swing from one year to the next, the introduction of 5-year farmers' averaging will help to smooth the results and the tax liabilities due.

There's nothing more important than our food supply. America is a country synonymous with wheat farms and orange trees. But according to McKinsey & Company, about a third of food produced is lost or wasted every year. Globally, that's a $940 billion economic hit. Inefficiencies in planting, harvesting, water use and trucking, as well as uncertainty about weather, pests, consumer demand and other intangibles contribute to the loss. On the consumer end, inadequate packaging and labeling can lead to waste and potentially life-threatening illness due to food-borne pathogens.

These are problems desperately in need of solutions and many of those solutions can be found in emerging technologies.

Big data is moving into agriculture in a big way. Need proof? Several well-known investors recently dropped a combined $40 million into Farmers Business Network, a data analytics startup. Venture capital has flooded the ag tech space, with investment increasing 80% annually since 2012, as investors realize big data can revolutionize the food chain from farm to table.

Sensors on fields and crops are starting to provide literally granular data points on soil conditions, as well as detailed info on wind, fertilizer requirements, water availability and pest infestations. GPS units on tractors, combines and trucks can help determine optimal usage of heavy equipment. Data analytics can help prevent spoilage by moving products faster and more efficiently. Unmanned aerial vehicles, or drones, can patrol fields and alert farmers to crop ripeness or potential problems. RFID-based traceability systems can provide a constant data stream on farm products as they move through the supply chain, from the farm to the compost or recycle bin. Individual plants can be monitored for nutrients and growth rates. Analytics looking forward and back assist in determining the best crops to plant, considering both sustainability and profitability. Agricultural technology can also help farmers hedge against losses and even out cash flow.

The software market for these sorts of precision farming tools (such as yield monitoring, field mapping, crop scouting and weather forecasting) is expected to grow 14% by 2022 in the United States alone. Researchers suggest the full-scale adoption of these technologies could mean an increase in farm productivity unseen since mechanization.

For consumers, packaging sensors detect gases emitted as food starts to spoil and verify packaging integrity and freshness. Algorithms can even help create a recipe out of whatever you have in the pantry. Several startups are building finger-sized scanners that tell the composition of food on your plate, from ingredients to nutrient content, by sending data to an app on your smartphone. These applications help not only health-conscience consumers but also those with chemical sensitivities or food allergies. Some projections say it could help reduce overall health care costs, too, as consumers are increasingly empowered to customize their nutrition and avoid potentially spoiled or contaminated foods.

Big data also holds enormous promise for urban farmers — people who are turning rooftops and abandoned lots into small farms. Lloyd Marino of Avetta Global, a big-data expert who has written about seed preservation, points out that, "Big data in conjunction with the Internet of Things can revolutionize farming, reduce scarcity and increase our nation's food supply in a dramatic fashion; we just have to institute policies that support farming modernization."

What's important now is to ensure that both the technology and data it generates are available to everyone. The U.S. Department of Agriculture should step up its support for the use of drones and other data systems for precision farming. Congress should get into the act and add a title to the Farm Bill that is due for reauthorization in 2018 that explicitly supports the widespread implementation of data and emerging tech to maximize efficient farming, save precious water, reduce unnecessary chemicals and decrease food waste and contamination.

Access to data for farmers, food handlers, grocers and the public shouldn't be cost prohibitive. Consumers and farmers both must trust the data, so how and why it's being collected should be transparent (and of course protected). We need smart industry standards and best practices for ag tech, new infrastructure such as smart roads to ensure we get the most from the technology, as well as an overhaul of communications infrastructure that wasn't designed for near-constant wireless input. Finally, research into farming robotics should be beefed up to develop robots that could respond to data for better, faster and more efficient production. In short, given the tremendous need for ag tech solutions we should all work to grow the use of precision farming and the application of wise data.

In the first year of the intervention, communities were randomly assigned to four groups. Within each community, researchers surveyed 20 randomly selected individual households and randomly assigned these households to sub-groups.

1. **Insurance Only:** In 50 communities, all farmers could purchase rainfall-indexed insurance, which makes payouts based on the amount of rainfall communities receive according to satellite data, at market price. The policies were designed by Ghana Agricultural Insurance Programme (GAIP) with input from Innovations for Poverty Action and then marketed by Community Based Marketers (CBMs). Of the 20 surveyed households, ten households received the insurance for free and ten served as a comparison group. In the final year of the study, 25 percent of the surveyed farmers received insurance coverage that offered a relatively high payout, while another 25 percent received coverage offering a lower payout.

2. **Insurance + Extension:** In addition to offers of insurance at market price, in 52 communities, randomly selected farmers received computer tablet-based interactive trainings and tailored recommendations on farming best practices for the cultivation of maize and legumes. Selected farmers received one-on-one extension services from Community Extension Agents (CEAs): paid, local workers that supplement traditional agricultural services from the Ministry of Agriculture. CEAs received a month of residential training and visited the selected farmers weekly to play specific videos and audio messages that were relevant to the current activities that farmers reported undertaking. Surveyed households were randomly assigned to one of four groups: free insurance, free insurance plus extension, extension only, and a comparison group. In the third year, researchers offered the extension services to the entire community in this group, although CEAs were required to deliver the messages to the extension treatment households.

3. **Insurance + Input Marketing and Delivery:** In addition to access to insurance, all farmers in 31 communities received the opportunity to buy commercial inorganic fertilizer, certified seeds and other agro-chemicals, and equipment at market price. During the first year of the study, farmers could purchase inputs at four points over the season, including at harvest when they are most likely to have cash on hand. Free delivery of these inputs occurred before or right at the start of planting time. Of the 20 surveyed households, ten households also received free insurance.

4. **Insurance + Extension + Input Marketing:** All farmers in 29 communities could purchase insurance and received access to inputs with delivery. In

addition, randomly selected farmers also received extension advice. The surveyed households were randomly assigned to one of four groups: free insurance plus access to inputs, free insurance plus extension plus access to inputs, extension plus access to inputs, and access to inputs only.

Additionally, in 108 randomly selected communities, 10 randomly selected farmers also received weather forecasts via SMS messages. Another 10 randomly selected households in the forecast communities served as comparison households. This allowed the researchers to evaluate the impacts of receiving weather forecasts both directly and indirectly.

6.3.1 Diversity in the Sector

Agrilyst's survey features respondents from a range technologies and operation types. Nearly half of respondents represent hydroponic farms, while 24% run soil-based operations, 15% aquaponics, 6% aeroponics, and 6% use a mixture of growing technologies.

The most dominant type of facility was glass or poly greenhouses (47%) followed by indoor vertical farms (30%), which generally operate in converted industrial buildings. Plastics hoop houses, container farms, and other structure types made up the remaining 23% of respondents.

Geographically, a clear majority (81%) of respondents were US-based, with 12% from Canada, and 7% from other countries.

One myth that the survey seeks to debunk is that indoor farming is an urban phenomenon. Urban farms receive more media attention and seem to draw more venture funding than other types of indoor farming; however, most of Agrilyst's survey respondents were based in rural locations (47%), while urban farms made up 43% of respondents and the rest were suburban.

"Indoor agriculture isn't equivalent to urban farming. This is a big misconception," the report states.

Farms make geographical decisions based on where they can maximize efficiency or near points of sale. A rural site may make the most sense for a tomato grower, for instance, because of lower costs of energy or where distribution centers are based.

As stated in a recent Technoserve report, in Colombia the average farm size has been diminishing due to a combination of demographic and economic factors, including inheritance traditions. Again, let's imagine that a rural family with three hectares of land has two or three children to inherit the property. This will lead to splitting the farm into ever-smaller production units that have an ever-harder time being economically viable.

But not many families are able to retain their children on their farms. Youngsters are continually overwhelmed with massive amounts of information via technology and social media, making migration to the cities look like the way to a "better future." Many young rural adults grew up in food-insecure families, and may equate coffee with poverty and hardship. Migration to more profitable and appealing industries is an increasing trend as emerging nations grow and transition from agriculture to manufacturing and service-based economies.

If those patterns continue, who will grow the coffee we all love 15 years from now? According to the report, "the youth are more likely to adopt new farming practices and technology and manage the farm more as a business," providing a unique opportunity for coffee to prosper along with them—but only if they take on the job to begin with. Therefore, coffee has to be an economically appealing and dignified profession that provides development opportunities for young women and men in rural communities, both on a personal and professional level.

Insufficient access to finance: Lack of access to formal credit for inputs and pre-harvest activities with commercial banks has led producers to commit their crops in advance to unscrupulous middlemen who charge them high interest rates and force them to deliver their coffee. Those coffees might not even be paid for at market price, and often times do not receive any economic premium for quality.

Absence of sufficient financial resources threatens the long-term viability of the farmers in many ways. Growers need the working capital to make the necessary investments in tree renovation, processing infrastructure, and production inputs that could allow them to reach optimal output, improve quality, reduce costs, and gain efficiencies over time. These circumstances impact productivity and price, the main pillars of farmer profitability.

Building resilience among coffee farmers requires collaborative industry efforts to create a strengthened sector that is better equipped to overcome environmental and economic pressures. The trade and the industry are already looking for ways to channel investments into their own supply chains to mitigate risks and assure the intended quantities of a quality product. Yet, it is important to acknowledge that given the magnitude of the challenges coffee faces, we alone cannot and will not solve them all.

Upgrading will also require investments in innovation, extension services, and improved market knowledge. Best agricultural practices are well known, so it is a matter of disseminating them effectively and efficiently. Technical assistance must include not only best-in-kind agricultural and sustainability practices, but also financial and managerial training that enables smallholders to position themselves as entrepreneurs capable of driving change and economic growth within their communities. Financial education for both producers and medium-size roasters is fundamental to manage supply and price risks, so all parties work with, and not against, the market and the tools it offers to better safeguard the interests of all stakeholders.

The fourth wave of specialty coffee will be driven by engaged, responsible supply chain management, where financial transparency and traceability are guaranteed, and a higher quality product will continue to deliver joy and energy for the consumer, and real benefits and pride for the growers.

Very often we have heard speakers stressing on "Doubling Farm's Income". Many people have stressed the use of more than one ways which would ensure profitable income for the farmers.

Though our Prime Minister has a vision to double farmer's income by 2022. But Mr. Yogendra stresses that if our farmers will adopt these five points, income of farmers can be double well before time.

Improving the production capacity of agriculture in developing countries through productivity increases is an important policy goal where agriculture represents an important sector in the economy. The agricultural sector provides livelihood directly and indirectly to a significant portion of the population of all developing countries, especially in rural areas, where poverty is more pronounced. Thus, a growing agricultural sector contributes to both overall growth and poverty alleviation.

Within the context of growth in food and agriculture, emphasis is placed on productivity because expansion of arable land is very limited in most countries due to physical lack of suitable land and/or because of environmental priorities. In addition, the difference between actual and technically feasible yields for most crops implies great potential for increasing food and agriculture production through improvements in productivity, even without further advances in technology.

Investment is of special interest as a limiting factor to agricultural production capacity and production because an alarming trend is being observed: public and private investment in agriculture has been declining. The decline in public investment is of particular concern because public investment in basic infrastructure, human capital formation and research and development (R&D) are necessary conditions for private investment. Public investments also promote technology adoption, stimulate complementary on-farm investment and input use and are needed for marketing the agricultural goods produced.

This paper provides an overview of the economic terminology, research findings, modelling techniques and their limitations that are used to link agricultural investment to output and productivity. After an introduction to the basic concepts and terminology of agricultural productivity and investment in the next section, the paper introduces three methodological approaches used to measure production growth, along with their advantages and disadvantages. In Section 1.4 findings from various agricultural growth studies are presented, focusing on the linkage between different types of agricultural investments and growth. Section 1.5 addresses the data needs and limitations in measuring the linkages between investments and growth in agriculture. These are of particular importance because data availability can limit the type of analysis done and consequently the questions that can be answered regarding agricultural growth and productivity. Concluding remarks are presented in the final section.

6.3.2 Investment

Investment is the change in fixed inputs used in a production process. In the most narrow definition, investment is the change in the physical capital stock, that is, physical inputs that have a useful life of one year or longer (land, equipment, machinery, storage facilities, livestock). However, Eisner (1985) estimated that less than 20 percent of total growth in the United States comes from physical capital formation, while Denison's (1967) estimates were 10 to 15 percent.

Economists recognize that, though difficult to measure, a comprehensive agricultural investment measure should include improvements in land, development of natural resources and development of human and social capital in addition to physical capital formation. Human capital is the stock of knowledge, expertise or management ability. Since it is directly influenced by educational, training and extension institutions, variables such as education level or extension contacts are often used as proxy measures. Public and private expenditures on R &D are often used to proxy the level of human capital as well. Coen and Eisner (1987) specifically include R&D, education and training as forms of human capital investment.

Social capital is the stock of personal relationships and knowledge of institutions that an individual or household has. This affects the individual's access to risk minimizing inputs like credit, insurance and land title. In other words, social capital measures the ability to utilize social networks and institutions. Status, gender and group affiliations are often used as proxies for social capital in economic studies. However, education and transportation, as well as the range of social institutions available, can also influence social capital.

6.4 ASYMMETRY OF INVESTMENT AND RISK

A key characteristic of investment is its irreversibility, often referred to as asymmetry (Nelson, Braden and Roh, 1989). Once investments are made, there are few other productive activities for which they can be used. Dixit and Pindyck (1994) formulate the problem of the irreversibility of investment under uncertainty as the decision to pay a sunk cost and in return receive an asset with a value that can fluctuate. They demonstrate that under uncertainty actual investment will always be less than the expected present value of investment, the difference being attributable to the irreversibility of industry specific investments.

Agro-climatic factors may exacerbate the asymmetry of agricultural investment, as is the case when the land is suitable only for a particular crop. Other forms of investment, such as tractors and farm machinery have few other alternative uses besides agriculture, while human and social capital particular to agriculture may not adapt well to other sectors. Contrast this with investments made in capital markets or even factories. The former can be moved around to the most profitable enterprise, while, in general, the latter can be modified to produce more profitable products. Due to this fixity of agricultural assets and the uncertainty it entails, farmers are often reluctant to invest in equipment, land improvements or human capital. Uncertainty may cause the level of investment to be "sub-optimal", resulting in deteriorating physical and human capital and mining of soil nutrients.

Drawing on fixed asset theory, Nelson, Braden and Roh (1989) hypothesize that it is more difficult to dispose of capital specific to agricultural production than to add to the stock of specialized capital. This implies that periods of disinvestment (through depreciation) will be greater than those of investment in agriculture. Thus, in any given year net agricultural investment is likely to be negative (depreciation is higher than gross investment). Because investment is irreversible, farmers only invest during years when profits are high and/or borrowing costs are low.

Rosenzweig and Binswanger (1993) find that agricultural investment behaviour of farmers reflects their risk aversion, with poorer farmers accepting lower returns in exchange for lower risk to smooth their consumption. The wealthy are less risk averse; they can afford to accept higher risk in seeking higher returns. Hence, they find that wealthier farmers, particularly those with larger farms and diversified incomes, have higher rates of farm investment on a per hectare basis. They suggest that consumption credit and/or crop insurance would increase the overall profitability of agricultural investments.

6.4.1 Public Expenditures and Investment in Agriculture

Public expenditures on agriculture include short-term costs as well as long-term investments. Investment in agriculture and forestry includes government expenditures directed to agricultural infrastructure, research and development and education and training. Data on the proportion of all central government expenditures spent on agriculture and forestry are incomplete, particularly for African countries. Comparisons between developed and developing countries reveal that there is greater variation among developing countries than industrial countries. In industrial countries in 1992, the range of expenditures was between 0.4 to 9.1 percent, with most countries clustered around 1.5 percent. For those developing countries reporting, agricultural expenditures were between 1.5 to 7.9 percent in Africa, 1.7 to 23 percent in Latin America and 0.20 to 19 percent in Asia. As a percentage of expenditures, agricultural expenditures generally declined from 1988 to 1993 in Africa, Eastern Europe and industrialized nations, declined for some Asian countries, increased for China and were mixed for Latin America.

Human capital development is a key component of public agricultural investment. Judd, Boyce and Evenson (1991) examined the role of public expenditures in agricultural research and extension on agricultural output. They show that between 1959 and 1980, real spending on research and extension programs increased by factors of four to seven and that research intensities more than tripled for the lowest income developing countries. They show a decrease in the disparity between countries over time. They estimate world agricultural research public-sector expenditures at US$7.4 thousand million and world public sector agricultural extension expenditures at US$3.4 thousand million (both in 1980 dollars). Africa had the smallest share of world research expenditures (5.7 percent) and human resources (5.5 percent), yet a larger share of world extension expenditures (14.8 percent) than Asia and the second largest world share of extension human resources (20.7 percent). Calculating public sector expenditures as a percent of agricultural product, Africa's expenditures are higher than those of South and Southeast Asia.

6.4.2 Measuring Agricultural Productivity

Models of production growth have been used to measure the change in output, to identify the relative contribution of different inputs to output growth and to identify the Solow residual or output growth not due to increases in inputs.

Three different types of economic models have been used to investigate production growth:

    i.   index numbers or growth accounting techniques,

    ii.   econometric estimation of production relationships and

    iii. nonparametric approaches.

Each approach can be used to measure aggregate agricultural output or TFP. Each approach has different data requirements, is suitable for addressing different questions and has strengths and weaknesses.

Growth accounting involves compiling detailed accounts of inputs and outputs, aggregating them into input and output indices to calculate a TFP index. The initial focus of growth accounting studies in the 1950s and 1960s was on partial measures of growth; only capital and labour were examined. However, growth accounting methods were unable to demonstrate much of a link between the amount of physical capital formation and output growth. Denison's (1967) growth accounting study of the 1950s and 1960s determined only 10 to 15 percent of growth could be accounted for by capital formation in non-residential plant and equipment. Nor did Bosworth (1982) find much of a role of reduced capital formation in the economic stagnation of the 1970s. Work by Abramovitz (1956), Solow (1957) and Kendrick (1973) "showed beyond reasonable doubt that the modern growth of the United States economy was in proportionate terms at least three-quarters due to increased efficiency in the use of productive inputs and not to the growth in the quantity of resource inputs per se ". This implied that quality of inputs matters more than quantity.

The failure of economics in the 1950s, 1960s and 1970s to find strong relationships between capital formation and economic growth was due in part to a narrow definition of capital formation and partly due to failure to control for other inputs. The unexplained growth was of the order of half the change in real output. Subsequent studies have tried to close this gap by including more inputs (fertilizer, pesticides, etc.), or finding ways to quantify inputs (human capital) for the analysis. The Solow residual has been referred to as efficiency, technological progress, economies of scale, or a "measure of our ignorance".

During the 1990s, there was a revival of interest in "new growth accounting" approaches, including endogenous growth models. The resurgence of interest in growth models has come in part from researchers incorporating omitted variables in their analysis, particularly measures of human capital, and new developments in the theory of growth. Hsieh (1998) developed a dual approach to computing the Solow residual using the growth in input prices rather than input quantities. Endogenous growth theory incorporates R&D as an intermediate input in the production process varieties model or views technological progress as improvements in the quality or cost of intermediate inputs quality ladder model. Obsolescence in technology differentiates the quality ladder model from the varieties model. Both models contain endogenously driven technological change and exogenous technological change.

6.5 FACTORS INFLUENCING GROWTH IN AGRICULTURE

Economists originally limited themselves to examining the roles of labour and physical capital in economic growth. The failure to adequately explain growth led them to examine the roles of other factors and to develop endogenous growth theory. Investment in infrastructure has been cited as an important source of growth in agriculture. However, Ferreira and Khatami (1996) claim that economic literature has not reached a consensus on the direction of causality between infrastructure and development. Nor can investment be viewed in isolation of policy reform which has been shown to be a vital stimulus of production; as have institutions. Public investment in forms of human capital: education, extension, training and technology research have also been shown to increase productivity.

Nelson (1964 and 1981) recognized that there are important interactions between capital formation, labour allocation, technical progress and productivity. This calls into question whether the growth due to physical capital can be separated from growth attributed to other inputs. Unless a production technology is a fixed Leontief process, there is always some degree of substitutability among categories of inputs. However, since inputs are not perfect substitutes, the lack of adequate investment can slow down production growth. Estimates of the elasticity of substitution in agriculture between hired labour and capital equipment vary from 0.32 in the short run to 1.78 percent in the long run.

Most measures of TFP incorporate inputs and physical capital, leaving human and social capital, technology, institutions, infrastructure and policy to "explain" growth in TFP. Social and human capital are the on-farm human elements that mediate how policy, technology, institutions and infrastructure affect input and physical capital use. Human capital directly affects whether and how technology will be adopted. Technology choice in turn, affects the inputs and physical capital used. That is, technology is embodied in the types of inputs and how they are used. Social capital affects access to physical capital (e.g.

land directly or through land titling and loans) and variable inputs (e.g. through credit or cooperatives).

In general, researchers have estimated TFP and then focused on how one or several of these factors might be driving its growth. Usually, they have done so using the change in TFP as a dependent variable in a regression with explanatory variables that represent measures of technology, human capital and policy (which are not easily quantifiable or assignable in constructing the production indices). In the following sections, policy is divided between budgetary policies that affect investment in R&D and infrastructure, political and economic policies and political stability.

6.5.1 Human Capital

Human capital directly influences agricultural productivity by affecting the way in which inputs are used and combined by farmers. Improvements in human capital affect acquisition, assimilation and implementation of information and technology. Human capital also affects one's ability to adapt technology to a particular situation or to changing needs.

Using an econometric approach, Nehru and Dhareshwar (1994) examined sources of TFP growth in 83 industrial and developing countries for the period 1960-1990. They found that human capital formation was three to four times more important than raw labour in explaining output growth. Using human capital as a separate variable, they found that the countries with the fastest growing economies have based their growth on factor accumulation (human capital, labour and physical capital), not growth in efficiency or technology.

6.5.2 Research and Technology Transfer

Research increases the set of available technologies, hence agricultural R&D expenditures are used as a proxy for agricultural technological change. However, the development of technology does not always result in its adoption. In some cases this may be because the technology being developed is not appropriate, that is, it does not meet the needs of agricultural producers. Hence, researchers focus on public expenditure as an explanatory variable in TFP growth. Additionally public research has been shown to lead private research.

While the returns to research are high, the technology is not always adopted. For example, high yield varieties (HYV) of wheat and rice have been introduced on less than one-third of the 423 million hectares planted to cereal grains in the Third World. Specifically, in Asia and the Middle East 36 percent of the grain area was HYV, 22 percent in Latin America and one percent in Africa. This implies there is much potential for increasing agricultural productivity using existing technology. However, the use of HYV requires increased use of fertilizer, but external debt in Latin America and poverty and inadequate water supply in Africa have made fertilizer use and hence HYV unprofitable. Jahnke, Kirschke and Lagemann (n.d.) also attributed low adoption of HYV in Africa to lack of appropriate technology development and few extension services directed to women. Additionally, nontraditional crops have rarely been the focus of improved varieties or technology and potential exists to develop them to increase agricultural production.

6.6 PUBLIC INVESTMENT AND POLICY

Public policy and budgetary decisions regarding infrastructure also have a profound effect on agricultural production. The financing aspects of public R&D and human capital development were discussed above, but both physical and institutional infrastructure affect the development and transfer of technology. For example, irrigation systems and roads may be required to make a technology profitable to implement. Reforms in pricing policy or the marketing system may be needed to provide incentives.

Using an econometric approach to estimate TFP for the United States dairy industry 1972-1992, Lachaal (1994) examined how protectionist policies in the form of direct subsidies to agriculture reduced productivity growth in the United States dairy industry. Lachaal showed that government subsidies encouraged using materials at the expense of feed and raised the cost of production by 1.8 percent for each 10 percent increase in subsidy. The subsidy policy was the source of technical inefficiency, creating biases that distorted factor usage.

6.6.1 Political Stability and Conflict

Another aspect of policy that can influence or hinder agricultural production is the political situation. In a study of the productivity growth of 83 industrial and developing countries between 1960 and 1990, Nehru and Dhareshwar (1994) found that the economies that perform the worst are those involved in wars (particularly civil wars) and those that have the most price distorting policies. They explore a variety of policy variables and find that apart from political stability and the initial endowments of a county, virtually no other policy variable is associated with growth.

The World Food Summit Plan of Action items 2 and 3 (1996b) recognize the role of government in providing an environment conducive to investment, through guarantee of rights and law as well as policies encouraging investment. Corruption is the extreme case where law enforcement breaks down and incentives are lacking. While long-standing institutionalized bribery can be seen as simply an added cost of doing business, pervasive corruption and violence increase risk and result in capital flight, disinvestment and jeopardize assistance.

6.6.2 Data Needs for Growth Models

In order to estimate any type of growth model, data are needed on agricultural output and inputs, including capital and labour. Comparable and consistent data are needed to make cross-country comparisons over time and space. The FAOSTAT under the World Agricultural Information Centre (WAICENT) is one of the most comprehensive agricultural databases created by FAO. FAOSTAT data are available by country and by year on agricultural production (crop and livestock), trade, land, economically active population in agricultural activity and means of production. The data on means of production include details of agricultural tractors, harvesters and threshers and milking machines in use, trade (export and import) portion of other agricultural machinery, fertilizers and pesticides. These data are generally expressed in quantity except for data on agricultural production and trade, which are in value.

Gross fixed capital formation under SEAFA will include:

1. Acquisitions less disposal of farm buildings and other structures, machinery and equipment, plantations, trees and livestock that can be used repeatedly or continually to produce fruit, rubber or milk;

2. Improvements to tangible assets including land; and

3. Costs associated with the transfer of ownership of non-produced assets.

Gross fixed capital formation specifically includes breeding stock, dairy cattle, sheep reared for wool and draught animals. Land improvements include reclamation of land by construction of dams, dikes or walls, drainage of marshes and flood control. Thus, SEAFA represents a vast improvement in the database on physical capital.

However, time series on gross fixed capital formation are not based on primary data collected in the field. Benchmark estimates of farm assets are extrapolated using quantity and price indices. Consumption of fixed capital is based on estimated useful life and estimated current value of the stock of fixed assets.

6.6.3 Data Issues

Regardless of the source of data, there are several data issues that are widely recognized in estimating growth models. Each is discussed below to highlight the importance and difficulty of developing consistent international data standards. However, developing data standards that will be in place a long time facilitate future researchers' ability to analyse and explain trends.

6.7 AGGREGATION

Griliches (1987) definition of productivity, "a ratio of some measure of output to some index of input use," highlights the vagaries of aggregating outputs and inputs. The physical units are simply not interchangeable unless converted to some common physical equivalent. However, monetary values are the most widely used method of aggregating both inputs and outputs, since monetary values can be summed together in a meaningful way and prices reflect the relative value of the items being aggregated.

6.7.1 Exchange Rates

In order to do cross country comparisons over time, data must be converted to a common real unit (i.e. adjusted for currency differences and inflation). However, an extensive literature outlines the problems associated with using official exchange rates to convert values to a common unit. The argument is that official exchange rates that do not reflect the actual currency values distort relative price relations. Purchasing power parity (PPP) has been used as an alternative. However, Antle (1983) argues that PPP has little relevance to agricultural output because it is based largely on non-agricultural goods and services and their use overstates agricultural production in developing countries. Another technique to avoid exchange rate distortions is to convert production to a common physical unit, such as "wheat units". Summers and Heston (1988) provide recommendations and develop a System of Real National Accounts that permits cross country comparisons.

6.7.2 Disaggregation

Insufficient disaggregation of inputs implies the inability to assign inputs to particular outputs. For example, the total amount of fertilizer or labour may be known, but how they are allocated among agricultural products may not. This is of particular importance when allocation of inputs is skewed to a minority of producers or crops such that reallocation could greatly improve total agricultural output.

Perhaps a greater problem exists with public expenditures and how to allocate them to agriculture. Rural development projects, for example, may have an agricultural component, but may not have an exclusively agricultural focus. Public education and training is rarely exclusively for agriculture, creating problems of how to allocate the expenditures to agriculture. Private education and training investments also are difficult to separate out an agricultural component.

6.8 MEASURING INPUTS AND OUTPUTS

A well-recognized problem is simply in measuring output. Kelly et al. (1995) estimate that data collection methods underestimate African agricultural production by up to 50 percent. This is because mixed cropping is common, crop by-products are not enumerated, crops are consumed at home or as inputs to other household production activities, or farmers have diversified into new products that are poorly enumerated in national surveys. On the input side, little data is available on small capital investments such as implements and land improvements, especially the value of family labour in land improvements.

6.8.1 Valuing Natural and Human Resources

Neither technology nor human capital can be quantified directly. Expenditures on research and extension have been successfully used as proxies for technology. Proxies for human capital are more problematic. Education level is generally available only at the national or regional level, not for the agricultural sector, thus is only a rough estimate of the level of agricultural human capital.

Measuring social capital generally requires a micro-data collection to develop a proxy. In village level studies, group affiliation or status has been used. At the country level, an aggregate proxy may be difficult to implement. One possibility is to use the percent of farms that are headed by males. Rural household income could be used as a proxy for relative wealth or status.

In official statistics, neither the value of natural resources nor the cost of environmental degradation are recognized in valuing land. Wolman (1985) estimates that ignoring these costs can be high. He reports agricultural productivity losses due to soil erosion up to 40 percent in the former USSR, 25 percent in the US, 30 percent in Haiti and 25 percent in Nigeria.

6.8.2 Lag Length and Dynamics of Investments

Another issue that affects data requirements, is exploring the time lag over which investments affect productivity. Capital investments by definition affect production in more than one year. The contribution of capital items to production diminishes or depreciates over time. In some cases the process may be linear, but in others the trajectory may be quite nonlinear or even discontinuous. Additionally, the process may be quite long. Chavas and Cox (1992) found that 30 years are required to fully capture the effects of public research expenditures in US agricultural productivity. This implies the need for extensive time series data to measure the effects of investments on productivity.

6.9 AGRICULTURE, ENVIRONMENT AND NATURAL RE-SOURCES

This is a Sectoral Committee that studies all matters related to agriculture, including crop and animal husbandry, livestock sale yards, county abattoirs, plant and animal disease control and fisheries; implementation of specific national government policies on natural resources and environmental conservation, including soil and water conservation and forestry and control of air pollution, noise pollution, other public nuisances and outdoor advertising. It is a permanent committee established by Standing orders of the Assembly.

6.9.1 Environment and Natural Resources Policy and Administration

Around three quarters of Cambodia's population depend on agriculture, forest products and fisheries for their livelihoods, so calculating the impact that any new development may have on the environment is important.

In the Constitution of Cambodia, the state has the responsibility to protect the environment and natural resources and establish a precise management plan for developing land, water, air, wind, geology, the ecological system, mines, energy, petrol and gas, rocks and sand, gems, forests and forestry products, wildlife, fish and aquatic resources.1 However, over the past decade, Cambodia's natural resources have faced different threats such as illegal logging, illegal and over-fishing, biodiversity depletion, and so on.

The Cambodian Government recognizes the importance of the country's natural resources. Cambodia's Rectangular Strategy - Phase III states that "the Royal Government of the Fifth Legislature will reinforce and broaden the management of natural resources to strike a balance between development and conservation, in particular, increase the contribution of natural resources to the development of the agriculture sector by ensuring:

- ◉ Green cover, forest and wildlife conservation
- ◉ The sustainability of fisheries resources
- ◉ The sustainability of the ecosystem, so that the quality of land and sustainability of water sources could be improved by focusing on the protection of biodiversity, wetlands and coastal areas."

6.9.2 Key Policies, Legal Framework and Institutions

Cambodia's main national development documents for governing environment and natural resources are:

- ◉ Rectangular Strategy Phase III
- ◉ National Strategic Development Plan 2014-2018
- ◉ National Forest Program 2010-2029
- ◉ Strategic Planning Framework for Fisheries Sector 2010-2019
- ◉ National Policy and Strategic Plan for Green Growth 2013-2030
- ◉ National REDD+ Roadmap.

In Cambodia, the main legislation regarding environmental protection and natural resource management is the 1996 Law on Environmental Protection and Natural Resources Management ("Environment Law"). Chapter III of the Environment Law requires that an environmental impact assessment (EIA) be conducted for projects likely to have an impact on the environment, whether they are public or privately funded.3 Click here for more details on the environmental impact assessment (EIA) in Cambodia.

As of late 2017 a new Environment and Natural Resources law is being drafted.4 A lawyer assisting with the drafting said the new law included nine chapters and more than 1,000 articles that deal with environmental issues and the protection of natural resources.5 The Ministry of Environment and outside experts had been working on the draft for two years.

The legal framework for managing forestry and fisheries are provided in the 2002 Forestry Law and 2006 Fisheries Law. The rights of communities to protect and manage forests and fisheries are recognized in Sub-Decree on Community Forestry Management and Sub-Decree on Community Fisheries Management respectively. Cambodia first defined its protected areas in a 1993 royal decree, and has since issued more detailed guidelines on how the country's protected areas must be managed in the form of the 2008 Protected Areas Law. The 2008 Protected Areas Law also provides a legal framework for managing the community protected areas.

Under the Protected Areas Law, the Ministry of Environment (MoE) is responsible for managing Cambodia's protected areas,6 while the Ministry of Agriculture, Forestry and Fisheries has jurisdiction on protected forest and fish sanctuaries. In 1999, Cambodia joined the Ramsar Convention, under which the Kingdom commits to the use of wetlands and water resources in a sustainable manner.7 To date, Cambodia has three Ramsar sites namely, Stung Treng, Boeung Chhamar and Koh Kapik, which are designated as globally important sites.8 See more on the types of state-protected area and protected areas pages.

The MoE is the main ministry responsible for implementing the Environment Law and associated regulations.9 Amongst its roles, the MoE is responsible for developing national and regional environmental management plans in order to identify key environmental challenges and the means to address them.10 Before issuing any decision or undertaking any activities related to the preservation, development, management or use of natural resources, concerned ministries must consult with the MoE.11 As environment and natural resources management are a cross-sectoral issue, the other institutions also play a role in environmental protection. More details can be found on the relevant ministries page. The government is currently drafting a Coastal Master Plan covering popular tourist beaches in Preah Sihanouk province. The aim is to enhance environmental protection and make the coastline attractive, clean and accessible for tourists. A strip 50 meters above the sea will be set aside for public use.

6.9.3 The Environment and Natural Resources

Humankind is still unsustainably consuming natural resources and services, beyond rates at which these resources can reproduce, regrow, and regenerate, exerting thereby increasing pressures on our climate, ecosystems, habitats, and biodiversity. Yet, the role of the environment and natural resources for development and wealth has remained poorly acknowledged in national and international policy designs so far.

A tipping or even turning point towards the design of sustainable national and international policies could be the United Nations 2030 Agenda for Sustainable Development and the UNFCCC Paris Agreement, calling for ambitious efforts to achieve sustainable development and strong climate change mitigation, respectively.

The RA "The Environment and Natural Resources" applies and develops models and practical tools to

⊙ Assess trade-offs between policy objectives,

- ⊙ Identify priorities for decision makers, and

- ⊙ Derive synergies and multi-purpose solutions,

contributing to the Kiel Institute´s mission of developing solutions for sustainable and inclusive prosperity in a globalized world.

6.10 ENVIRONMENT AND NATURAL RESOURCE MANAGE-MENT (NRM)

Keeping agriculture sustainable and profitable is essential to the longevity and prosperity of our sector and the economy. QFF works closely with its industry members and their farmers to encourage sustainable farming systems that minimising cost inputs and benefit the environment.

Qff's environment and nrm agenda continues to be both proactive (in terms of promoting best practices) and reactive (in responding to government and community concerns). Environment and nrm policies are a priority in qff advocacy. By working closely with government, qff ensures that a balanced and responsible approach is taken in the long-term development of queensland agriculture. Qff maintains productive partnerships with government and the community to ensure the delivery of mutually beneficial, profitable and environmentally resilient outcomes.

6.10.1 Agricultural Production Trends in India: An Overview

Prior to Independence

It may be pointed out that during the period 1901 to 1947, agricultural production declined.

The population rose by 38 per cent while the increase in cultivated area was to the extent of 18 percent. The annual output of food grains and pulses remained almost constant.

Pandse has made a special study of the yield of principal crops in India for the period between 1910-11 and 1945-46 and concluded that the yield per acre of cereals did not show any consistent decline or increase but there was a positive increase in the yield per acre of commercial crops and food-grains. He did not agree with the belief that there had been deterioration in fertility or in the standards of agriculture.

Post Independence Period

The process of decline in productivity has continued in the post-independence period, as compared to the pre-1939 period. The average yield of cereals per acre during 1946-47 to 1949-50 had declined from 619 to 565 lbs. Rangnekar found that the volume of output in India declined from 0.9 metric tones in 1938-39 to 0.86 metric tones per hectare in 1951. Similar conclusion were reached by studies undertaken by ICAR and the Grow More Food Enquiries.

With the introduction of economic planning in 1951 and with the special emphasis on agriculture development, particularly after 1962, stagnant of agriculture was reversed as:

1. There was a steady rise in average yield per hectare.

2. There was a Steady rise in area under cultivation.

3. Due to increase in area and increase in yield per hectare, total production of the crops recorded a rising trend.

Trends in Food-Grains Production

The increase in agricultural production has an important impact on the economic development of a country.

It reveals, that in the last fifty two years food-grains production has increased by about more than three times. The increase in the production of rice was four times while it was over nine times in respect of wheat. Here, it is worth noting that there exists wide variations in the production of food-grains.

Regarding trend of rice and wheat, it was 276 lakh tones and 86 lakh tones in first plan which increased to 441 lakh tones and 218 lakh tones in fourth plan and further to 798 lakh tones and 628 lakh tones respectively in 1995- 96. During 2002-03, production of total food-grains 1742 lakh tones, pulses 111 lakh tones, rice 727 million tones and wheat 651 million tones.

Non-Food Grains

The trends in non-food grains production in India after the introduction of economic planning is shown in table 3.

The production of sugarcane has shown upward trend as it was 570 lakh tones in 1950-51, 1520 lakh tones in 1980-81, and 2540 lakh tones in 1991-92. At the close of eighth five year plan, its production was registered at 2829 lakh tones. In case of oil seeds, production was 186 lakh tones in 1990-91 against 50 lakh tones in 1950-51, and 101 lakh tones in 1980-81.

During 1995- 96, the production of oilseeds was recovered 224 lakh tones. Similarly, production of cotton, jute, sugarcane was 87,103 and 2816 lakh tones and production of oilseeds was 151 million tones in 2002-03.

6.11 FACTORS OF PRODUCTION - THE AGRICULTURE ECONOMIC

The factors of production are resources that are the building blocks of the economy; they are what people use to produce goods and services. Economists divide the factors of production into four categories: land, labor, capital, and entrepreneurship.

The first factor of production is land, but this includes any natural resource used to produce goods and services. This includes not just land, but anything that comes from the land. Some common land or natural resources are water, oil, copper, natural gas, coal, and forests. Land resources are the raw materials in the production process. These resources can be renewable, such as forests, or nonrenewable such as oil or natural gas. The income that resource owners earn in return for land resources is called rent.

The second factor of production is labor. Labor is the effort that people contribute to the production of goods and services. Labor resources include the work done by the waiter who brings your food at a local restaurant as well as the engineer who designed the bus that transports you to school. It includes an artist's creation of a painting as well as the work of the pilot flying the airplane overhead. If you have ever been paid for a job, you have contributed labor resources to the production of goods or services. The income earned by labor resources is called wages and is the largest source of income for most people.

The third factor of production is capital. Think of capital as the machinery, tools and buildings humans use to produce goods and services. Some common examples of capital include hammers, forklifts, conveyer belts, computers, and delivery vans. Capital differs based on the worker and the type of work being done. For example, a doctor may use a stethoscope and an examination room to provide medical services. Your teacher may use textbooks, desks, and a whiteboard to produce education services. The income earned by owners of capital resources is interest.

The fourth factor of production is entrepreneurship. An entrepreneur is a person who combines the other factors of production - land, labor, and capital - to earn a profit. The most successful entrepreneurs are innovators who find new ways produce goods and services or who develop new goods and services to bring to market. Without the entrepreneur combining land, labor, and capital in new ways, many of the innovations we see around us would not exist. Think of the entrepreneurship of Henry Ford or Bill Gates. Entrepreneurs are a vital engine of economic growth helping to build some of the largest firms in the world as well as some of the small businesses in your neighborhood. Entrepreneurs thrive in economies where they have the freedom to start businesses and buy resources freely. The payment to entrepreneurship is profit.

You will notice that I did not include money as a factor of production. You might ask, isn't money a type of capital? Money is not capital as economists define capital because it is not a productive resource. While money can be used to buy capital, it is the capital good (things such as machinery and tools) that is used to produce goods and services. When was the last time you saw a carpenter pounding a nail with a five dollar bill or a warehouse foreman lifting a pallet with a 20 dollar bill? Money merely facilitates trade, but it is not in itself a productive resource.

Remember, goods and services are scarce because the factors of production used to produce them are scarce. In case you have forgotten, scarcity is described as limited quantities of resources to meet unlimited wants. Consider a pair of denim blue jeans. The denim is made of cotton, grown on the land. The land and water used to grow the cotton is limited and could have been used to grow a variety of different crops. The workers who cut and sewed the denim in the factory are limited labor resources who could have been producing other goods or services in the economy. The machines and the factory used to produce the jeans are limited capital resources that could have been used to produce other goods. This scarcity of resources means that producing some goods and services leaves other goods and services unproduced.

It's time to test your knowledge with a little game I like to call, Name That Resource. I will say the name of an item and you will identify it as one of the four possible resources that form the factors of production: land, labor, capital, or entrepreneurship.

- ◉ Coal... land
- ◉ Forklift... capital
- ◉ Factory... capital
- ◉ Oil... land
- ◉ Michael Dell... entrepreneur

It's time to wrap things up, but before we go, always remember that the four factors of production - land, labor, capital, and entrepreneurship - are scarce resources that form the building blocks of the economy.

6.11.1 Production Definition in Economics

Production in ordinary sense means creation of a commodity. We say the carpenter has produced the chair. But in Economics it is a wrong view. The carpenter has given shape to the wood which is a free gift of nature as a result of which it has become more useful to us than before. He has strictly speaking, created additional utility. So production in Economics means creation of new utility. Man takes the things given by nature and simply gives it a new form so that it becomes more useful to us than before.

(a) By changing the form of an object of nature, viz., iron ore into steel, wood into furniture. It is known as form utility.

(b) By changing the place, i.e., transferring a thing from the place of abundance to the place of scarcity. It is called place utility.

(c) Utility may be increased by transferring a thing from one time to another, i.e., when it is relatively abundant to a time when it is scarce. It is what is known as place utility.

Production requires co-operation of certain factors. These are known as agents of production. Broadly, there are four such agents, namely, land, labour, capital and organisation. Land includes both manual and intellectual labour. Capital is produced means of production.

Organisation is a broad term. It is the factor that faces all the challenges and hazards of production. It pilots the ships of production unit through storm and strain. Factors of production may again be classified into two categories- fixed factors and variable factors.

The former include those factors whose quantity cannot be changed in the short run such as capital goods. The cost incurred for such factors is known as fixed cost or supple-mentary cost. There are some other factors the quantity of which must vary with the level of output, e.g., raw materials cost, casual labour costs, etc. In the long run all factors are likely to be variable.

6.11.2 Different Types of Production

Since the purpose of any economic activity is the satisfaction of human wants, any activity which helps to satisfy wants is defined as production. In order to survive man must consume; in order to consume he must produce.

In fact, consumption needs determine production plans, and the actual production satisfies those original consumption needs. This, in short, is the economic cycle.

To quote Adam Smith:

"Consumption is the sole end purpose of all production; and the interest of the producer ought to be attended to, only so far as it may be necessary for promoting that of the consumer".

Since the primary purpose of economic activity is to produce utility for individuals, we count as production during a time period all activity which either creates utility during the period or which increases ability of the society to create utility in the future. Simply put, the making of goods and providing services is known as production.

Those providing services ensure that wants are met and goods are sold in the form, at the time, and when and where they are required. A dentist is just as productive in economic terms as a farmer, as both fulfil needs and get paid for doing so. Thus, any paid employment which arises from the supply of raw materials to the consumption of a good or service may be considered as productive.

Direct production refers to a worker supplying his own needs, i.e., self-sufficient, subsistence farming. In modern society, nearly all production is indirect with people producing goods and services for others.

There are three aspects (components) of the production process, viz., inputs (or factors of production), output (saleable goods having utility or want-satisfying capacity) and technology (the art or method of production). Services are also considered as being produced. Inputs are the beginning of the production process and output is the end of the process. Technology lies at the intermediate stage of the whole process.

The producing unit is either the agricultural farm or the (industrial) factory. Business firms are important components (units) of the economic system. They are artificial entities created by individuals for the purpose of organising and facilitating production. It is a technical unit in which inputs are converted into output.

The essential characteristics of the business firm is that it purchases factors of production such as land, labour, capital, intermediate goods, and raw material from households and other business firms and transforms those resources into different goods or services which it sells to its customers, other business firms and various units of the government as also to foreign countries.

6.11.3 Types of Production

For general purposes, it is necessary to classify production into three main groups:

1. **Primary production:** Primary production is carried out by 'extractive' industries like agriculture, forestry, mining and oil extraction. These industries are engaged in such activities as extracting the gifts of Nature from beneath the earth's surface and from the oceans. Primary activities refer to such things as extraction of raw materials from the earth's surface, e.g., coal mining or pisiculture (fishing). In advanced countries, the primary sector is providing less employment because machinery is replacing man power.

2. **Secondary production:** This includes production in manufacturing industry, viz., turning out semi-finished and finished goods from raw materials and intermediate goods - conversion of flour into bread or iron ore into finished steel. These activities are generally described as manufacturing and construction industries, such as the manufacture of cars, furnishing, clothing and chemicals, as also engineering and building. In short, secondary production is concerned with conversion of raw materials into finished products, e.g., manufacturing motor cars, shirts, medicines, food, etc.

3. **Tertiary production:** Industries in the tertiary sector produce all those services which enable the finished goods to be put in the hands of consumers. In fact, these services are supplied to the firms in all types of industries and directly to consumers. Examples cover distributive traders, banking, insurance, transport and communications. Government services, such as law, administration, education, health and defence, are also included.

6.11.4 Factors of Production

Production in ordinary sense means creation of a commodity. We say the carpenter has produced the chair. But, in Economics, it is a wrong view. The carpenter has given shape to the wood which is a free gift of nature as a result of which it has become more useful to us than before. He has, strictly speaking, created additional utility. So, production in Economics means creation of new utility. Man takes a thing given by nature and simply gives it a new form so that it becomes more useful to us than before.

Man may create additional utility in at least three ways:

(a)   By changing the form of an object of nature, viz., iron ore into steel, wood into furniture. It is called form utility,

(b)   Utility can also be created by changing the place, i.e., transferring a thing from the place of abundance to the place of scarcity. It is called place utility,

(c)   Finally, utility may be increased by transferring a thing from one time to another - when it is relatively abundant to a time when it is scarce. It is what is known as time utility.

Production requires cooperation of certain factors. These are known as agents of production. They are also called economic resources or inputs. Broadly, there are four such agents, namely, land, labour, capital and organisation. Labour includes both natural and intellectual labour. Capital is produced means of production. Organisation is a broad term. It is the factor that faces all the challenges and hazards of production. It pilots the ship of production unit through storm and strain. Factors of production may again be classified into two categories - fixed factors and variable factors. The former include those factors whose quality cannot be changed in the short-run such as capital goods. The cost incurred for such factors is known as fixed cost or supplementary cost. There are some other factors the quantity of which must vary with the level of output - e.g., raw materials cost, casual labour costs, etc. In the long-run all factors are likely to be variable.

All productive processes require the above four factors in varying proportions

Land

Land in the economic sense is the natural resources of this planet. It is not only land itself but also what lies under the land (like coal and gold), what grows naturally on top of the land (forests, wild animals), what is over the land (like the air) and what is around the land in the seas and oceans and under the seas and oceans (like fish and oil). Only one major resource is for the most part free - the air we breathe. The rest are scarce, because there are not enough natural resources in the world to satisfy the demands of consumers and producers. A natural resource covers all 'free gifts of nature', e.g., earth, trees, flat land, sea, rivers, etc. Land can be bought or rented, but it is necessary before production can be started. The owners of land receive rent for its use.

Labour

Labour is the human input into the production process.

Two important points are to be remembered about labour as a resource:

(1)   Just because a person has not got a paid job, it does not mean that he or she does not produce goods and services;

(2)   Not all labour is of the same quality.

Some workers are more productive than others because of the education, training and experience they have received. This is called human capital. The greater the human capital of a worker the more productive he or she will be workers of every type in every kind of activity from surgeons to shop assistants represents human resources.

Different jobs require different levels of strength, skill, education and different types of responsibility. The bigger the organisation then usually the wider is the variety of human work required. Labour is 'owned' by individuals who sell it to firms and receive wages/salaries in return.

Capital

Capital is a man-made resource, e.g., machinery, a lorry or a robot. It is used to make consumer goods and services. Without capital, there would be no production. Capital goods are those things which human beings have produced, in order to create other goods and services. Capital goods are the creators of other goods and the demand for capital goods is a derived demand.

A modern industrialised economy possesses a large amount of capital and it is continually increasing. Increases to the capital stock of a nation are called investment. This capital is sometimes called physical capital or non-human capital in order to distinguish it from human capital.

Usually capital and labour are combined. Capital lasts a long time but eventually needs replacing. When its value declines with age, it is said to be 'depreciating'. Some industries are labelled as 'capital-intensive' in that they have few labour costs and rely heavily on automated machinery, e.g., the chemical industry. The money borrowed to provide capital is paid interest.

Enterprise

It is another human resource. This factor refers to the organising, planning and risk-taking by the owner of a business. An entrepreneur is an individual who risks his own resources (money in most cases) in a business venture; and who organises the business - i.e., organises the other three factors of production. An entrepreneur is a special type of worker.

Many economists agree that the entrepreneur should be classed as part of the factor 'labour'. He receives profit for his work, and is called the entrepreneur. However, in modern economies large businesses are seldom owned by one person; instead they are owned by many shareholders and controlled by a Board of Directors.

Not all enterprises aim to make a profit. Also some nationalised industries operate in order to provide a service rather than to make a profit.

Factor Mobility

One characteristic of productive factors of considerable importance is their mobility. Factor mobility refers to the extent to which (or ease with which) factors of production can move from one occupation to another or one region (geographic area) to another in response to changes in market conditions (factor prices).

Factor mobility is important for two reasons

(1)   Firstly, factor mobility influences business decision-making, particularly when there is some change in consumer demand conditions calling for adjustment.

(2)   Secondly, changes in the character of national product (GNP) largely depend on the degree of factor mobility.

Such mobility is of following two types:

1. Geographic Mobility

The simplest aspect of mobility is geographical. It refers to movement of a factor from one area (region) to another.

It may be observed that some factors like raw materials and small items of equipment can move freely from one place to another (or one country to another if, of course, there is no government restriction) in response to needs, while others are much less able to do that. In the extreme case of land mobility is, of course, zero. Labour occupies an intermediate position. People can, at least in principle, settle almost anywhere, though there is often strong social resistance to uprooting.

2. Occupational Mobility

A second aspect of factor mobility is occupational. It is concerned with the mobility of a factor from one occupation (industry) to another or from one use to another. It is often important (and interesting) to know how easily a factor such as labour can shift from, say, motor vehicles to bicycle production or land from growing wheat to grazing cattle. However, two generalizations may be made in this context.

Factors mobility depends on time and cost:

(a) Time

Firstly, factors tend to become mobile in the long run than in the short run.

(b) Cost

Secondly, there is frequently a cost involved in moving.

Let us consider labour as an example. If industries in one part of the country decline, it is unlikely that people will immediately leave the area. However, with the passage of time more and more workers will tend to move. Moreover, various costs are incurred in such a move. Houses and flats are needed in areas of expansion. There are psychological costs, too, such as disturbances to family life.

One final point may be noted. Sometimes, the government deliberately erects certain barriers to mobility. For example, certain jobs are reserved for special categories of people (such as retired army personnel, scheduled caste and scheduled tribe candidates).

6.12 THEORY OF PRODUCTION

Theory of production, in economics, an effort to explain the principles by which a business firm decides how much of each commodity that it sells (its "outputs" or "products") it will produce, and how much of each kind of labour, raw material, fixed capital good, etc., that it employs (its "inputs" or "factors of production") it will use. The theory involves some of the most fundamental principles of economics. These include the relationship between the prices of commodities and the prices (or wages or rents) of the productive factors used to produce them and also the relationships between the prices of commodities and productive factors, on the one hand, and the quantities of these commodities and productive factors that are produced or used, on the other.

The various decisions a business enterprise makes about its productive activities can be classified into three layers of increasing complexity. The first layer includes decisions about methods of producing a given quantity of the output in a plant of given size and equipment. It involves the problem of what is called short-run cost minimization. The second layer, including the determination of the most profitable quantities of products to produce in any given plant, deals with what is called short-run profit maximization. The third layer, concerning the determination of the most profitable size and equipment of plant, relates to what is called long-run profit maximization.

6.12.1 Minimization of Short-Run Costs

The cost of production is simply the sum of the costs of all of the various factors. It can be written:

The principles involved in selecting the cheapest combination of variable factors can be seen in terms of a simple example. If a firm manufactures gold necklace chains in such a way that there are only two variable factors, labour (specifically, goldsmith-hours) and gold wire, the production function for such a firm will be $y = f(x1, x2; k)$, in which the symbol k is included simply as a reminder that the number of chains producible by x1 feet of gold wire and x2 goldsmith-hours depends on the amount of machinery and other fixed capital available. Since there are only two variable factors, this production function can be portrayed graphically in a figure known as an isoquant diagram (Figure 1). In the graph, goldsmith-hours per month are plotted horizontally and the number of feet of gold wire used per month vertically. Each of the curved lines, called an isoquant, will then represent a certain number of necklace chains produced. The data displayed show that 100 goldsmith-hours plus 900 feet of gold wire can produce 200 necklace chains. But there are other combinations of variable inputs that could also produce 200 necklace chains per month. If the goldsmiths work more carefully and slowly, they can produce 200 chains from 850 feet of wire; but to produce so many chains more goldsmith-hours will be required, perhaps 130. The isoquant labelled "200" shows all the combinations of the variable inputs that will just suffice to produce 200 chains. The other two isoquants shown are interpreted similarly. It is obvious that many more isoquants, in principle an infinite number, could also be drawn. This diagram is a graphic display of the relationships expressed in the production function.

6.12.2 Marginal Cost

Two other concepts now become important. The average variable cost, written AVC(y), is the variable cost per unit of output. Algebraically, AVC(y) = VC(y)/y. The marginal variable cost, or simply marginal cost[MC(y)] is, roughly, the increase in variable cost incurred when output is increased by one unit; i.e., MC(y) = VC(y + 1) - VC(y). Though for theoretical purposes a more precise definition can be obtained by regarding VC(y) as a continuous function of output, this is not necessary in the present case.

The usual behaviour of average and marginal variable costs in response to changes in the level of output from a given fixed plant is shown in Figure 3. In this figure costs (in dollars per unit) are measured vertically and output (in units per year) is shown horizontally. The figure is drawn for some particular fixed plant, and it can be seen that average costs are fairly high for very low levels of output relative to the size of the plant, largely because there is not enough work to keep a well-balanced work force fully occupied. People are either idle much of the time or shifting, expensively, from job to job. As output increases from a low level, average costs decline to a low plateau. But as the capacity of the plant is approached, the inefficiencies incident on plant congestion force average costs up quite rapidly. Overtime may be incurred, outmoded equipment and inexperienced hands may be called into use, there may not be time to take machinery off the line for routine maintenance; or minor breakdowns and delays may disrupt schedules seriously because of inadequate slack and reserves. Thus the AVC curve has the flat-bottomed U-shape shown. The MC curve, as might be expected, falls faster and rises more rapidly than the AVC curve.

6.12.3 Maximization of Short-Run Profits

The average and marginal cost curves just deduced are the keys to the solution of the second-level problem, the determination of the most profitable level of output to produce in a given plant. The only additional datum needed is the price of the product, say p0.The most profitable amount of output may be found by using these data. If the marginal cost of any given output (y) is less than the price, sales revenues will increase more than costs if output is increased by one unit (or even a few more); and profits will rise. Contrariwise, if the marginal cost is greater than the price, profits will be increased by cutting back output by at least one unit. It then follows that the output that maximizes profits is the one for which MC(y) = p0. This is the second basic finding: in response to any price the profit-maximizing firm will produce and offer the quantity for which the marginal cost equals that price. Such a conclusion is shown in Figure 3. In response to the price, p0, shown, the firm will offer the quantity y* given by the value of y for which the ordinate of the MC curve equals the price. If a denotes the corresponding average variable cost, net revenue per unit will be equal to p0 - a, and the total excess of revenues over variable costs will be y*(p0- a), which is represented graphically by the shaded rectangle in the figure.

6.13 MARGINAL COST AND PRICE

The conclusion that marginal cost tends to equal price is important in that it shows how the quantity of output produced by a firm is influenced by the market price. If the market price is lower than the lowest point on the average variable cost curve, the firm will "cut its losses" by not producing anything. At any higher market price, the firm will produce the quantity for which marginal cost equals that price. Thus the quantity that the firm will produce in response to any price can be found in Figure 3 by reading the marginal cost curve, and for this reason the marginal cost curve is said to be the short-run supply curve for the firm.

The short-run supply curve for a product-that is, the total amount that all the firms producing it will produce in response to any market price-follows immediately, and is seen to be the sum of the short-run supply curves (or marginal cost curves, except when the price is below the bottoms of the average variable cost curves for some firms) of all the firms in the industry. This curve is of fundamental importance for economic analysis, for together with the demand curve for the product it determines the market price of the commodity and the amount that will be produced and purchased.

One pitfall must, however, be noted. In the demonstration of the supply curves for the firms, and hence of the industry, it was assumed that factor prices were fixed. Though this is fair enough for a single firm, the fact is that if all firms together attempt to increase their outputs in response to an increase in the price of the product, they are likely to bid up the prices of some or all of the factors of production that they use. In that event the product supply curve as calculated will overstate the increase in output that will be elicited by an increase in price. A more sophisticated type of supply curve, incorporating induced changes in factor prices, is therefore necessary. Such curves are discussed in the standard literature of this subject.

6.13.1 Marginal Product

It is now possible to derive the relationship between product prices and factor prices, which is the basis of the theory of income distribution. To this end, the marginal product of a factor is defined as the amount that output would be increased if one more unit of the factor were employed, all other circumstances remaining the same. Algebraically, it may be expressed as the difference between the product of a given amount of the factor and the product when that factor is increased by an additional unit. Thus if $MP1(x1)$ denotes the marginal product of factor 1 when $x1$ units are employed, then $MP1(x1) = f(x1 + 1, x2,..., xn; k) - f(x1, x2..., xn; k)$. The marginal products are closely related to the marginal rates of substitution previously defined. If an additional unit of factor 1 will increase output by $f1$ units, for example, then one more unit of output can be obtained by employing $1/f1$ more units of factor 1. Similarly, if the marginal product of factor 2 is $f2$, then output will fall by one unit if the use of factor 2 is reduced by $1/f2$ units. Thus output will remain unchanged, to a good approximation, if $1/f1$ units of factor 1 are used to replace $1/f2$ units of factor 2. The marginal rate of substitution is therefore $f2/f1$, or the ratio of the marginal products of the two factors. It has already been shown that the marginal rate of substitution also equals the ratio of the prices of the factors, and it therefore follows that the prices (or wages) of the factors are proportional to their marginal products.

This is one of the most significant theoretical findings in economics. To restate it briefly: factors of production are paid in proportion to their marginal products. This is not a question of social equity but merely a consequence of the efforts of businessmen to produce as cheaply as possible.

Further, the marginal products of the factors are closely related to marginal costs and, therefore, to product prices. For if one more unit of factor 1 is employed, output will be increased by $MP1(x1)$ units and variable cost by $p1$; so the marginal cost of additional units produced will be $p1/MP1(x1)$. Similarly, if additional output is obtained by employing an additional unit of factor 2, the marginal cost will be $p2/MP2(x2)$. But, as shown above, these two numbers are the same; whichever factor i is used to increase output, the marginal cost will be $pi/MPi(xi)$ and, furthermore, the firm will choose its output level so that the marginal cost will be equal to the price, $p0$.

Therefore it has been established that $p1 = p0MP1(x1)$, $p2 = p0MP2(x2)$,..., or the price of each factor is the price of the product multiplied by its marginal product, which is the value of its marginal product. This, also, is a fundamental theorem of income distribution and one of the most significant theorems in economics. Its logic can be perceived directly. If the equality is violated for any factor, the businessman can increase his profits either by hiring units of the factor or by laying them off until the equality is satisfied, and presumably the businessman will do so.

The theory of production decisions in the short run, as just outlined, leads to two conclusions (of fundamental importance throughout the field of economics) about the responses of business firms to the market prices of the commodities they produce and the factors of production they buy or hire: (1) the firm will produce the quantity of its product for which the marginal cost is equal to the market price and (2) it will purchase or hire factors of production in such quantities that the price of the commodity produced multiplied by the marginal product of the factor will be equal to the cost of a unit of the factor. The first explains the supply curves of the commodities produced in an economy. Though the conclusions were deduced within the context of a firm that uses two factors of production, they are clearly applicable in general.

7

MARKETING

7.1 OBJECTIVES

After studying this chapter you will be able to understand:

- Marketing Research and analysis.
- Marketing Plan.
- Marketing Management.
- Budgets and Managerial Tools.

7.2 INTRODUCTION

Agricultural marketing is inferred to cover the services involved in moving an agricultural product from the farm to the consumer. It is also the planning, organizing, directing and handling of agricultural produce in such a way as to satisfy the farmer, producer and the consumer. Numerous interconnected activities are involved in doing this, such as planning production, growing and harvesting, grading, packing and packaging, transport, storage, agro- and food processing, distribution, advertisingand sale.

Effectively, the term encompasses the entire range of supply chain operations. However, it's key function is to help direct these services, by providing competent and able market information, thereby linking the other operations into an integrated service with targeted outcomes.

7.3 ACTIVITIES AND FUNCTIONS

Marketing management therefore encompasses a wide variety of functions and activities, although the marketing department itself may be responsible for only a subset of these. Regardless of the organizational unit of the firm responsible for managing them, marketing management functions and activities include the following:

7.3.1 Marketing Research and Analysis

In order to make fact-based decisions regarding marketing strategy and design effective, cost-efficient implementation programs, firms must possess a detailed, objective understanding of their own business and the market in which they operate. In analyzing these issues, the discipline of marketing management often overlaps with the related discipline of strategic planning.

Traditionally, marketing analysis was structured into three areas: Customer analysis, Company analysis, and Competitor analysis (so-called "3Cs" analysis). More recently, it has become fashionable in some marketing circles to divide these further into certain five "Cs": Customer analysis, Company analysis, Collaborator analysis, Competitor analysis, and analysis of the industry Context.

The focus of customer analysis is to develop a scheme for market segmentation, breaking down the market into various constituent groups of customers, which are called customer segments or market segments. Marketing managers work to develop detailed profiles of each segment, focusing on any number of variables that may differ among the segments: demographic, psychographic, geographic, behavioral, needs-benefit, and other factors may all be examined. Marketers also attempt to track these segments' perceptions of the various products in the market using tools such as perceptual mapping.

In company analysis, marketers focus on understanding the company's cost structure and cost position relative to competitors, as well as working to identify a firm's core competencies and other competitively distinct company resources. Marketing managers may also work with the accounting department to analyze the profits the firm is generating from various product lines and customer accounts. The company may also conduct periodic brand audits to assess the strength of its brands and sources of brand equity.

The firm's collaborators may also be profiled, which may include various suppliers, distributors and other channel partners, joint venture partners, and others. An analysis of complementary products may also be performed if such products exist.

Marketing management employs various tools from economics and competitive strategy to analyze the industry context in which the firm operates. These include Porter's five forces, analysis of strategic groups of competitors, value chain analysis and others. Depending on the industry, the regulatory context may also be important to examine in detail.

In Competitor analysis, marketers build detailed profiles of each competitor in the market, focusing especially on their relative competitive strengths and weaknesses using SWOT analysis. Marketing managers will examine each competitor's cost structure, sources of profits, resources and competencies, competitive positioning and product differentiation, degree of vertical integration, historical responses to industry developments, and other factors.

Marketing management often finds it necessary to invest in research to collect the data required to perform accurate marketing analysis. As such, they often conduct market research (alternately marketing research) to obtain this information. Marketers employ a variety of techniques to conduct market research, but some of the more common include:

- Qualitative marketing research, such as focus groups
- Quantitative marketing research, such as statistical surveys
- Experimental techniques such as test markets
- Observational techniques such as ethnographic (on-site) observation

Marketing managers may also design and oversee various environmental scanning and competitive intelligence processes to help identify trends and inform the company's marketing analysis.

7.3.2 Marketing Strategy

A marketing strategy is a process that can allow an organization to concentrate its limited resources on the greatest opportunities to increase sales and achieve a sustainable competitive advantage. A marketing strategy should be centered around the key concept that customer satisfaction is the main goal.

Key Components of Marketing Strategy

A marketing strategy is most effective when it is an integral component of firm strategy, defining how the organization will successfully engage customers, prospects, and competitors in the market arena. Corporate strategies, corporate missions, and corporate goals. As the customer constitutes the source of a company's revenue, marketing strategy is closely linked with sales. A key component of marketing strategy is often to keep marketing in line with a company's overarching mission statement.

Basic theory:

1. Target Audience
2. Proposition/Key Element
3. Implementation
4. The Five D's

7.3.3 Tactics and Actions

A marketing strategy can serve as the foundation of a marketing plan. A marketing plan contains a set of specific actions required to successfully implement a marketing strategy. For example: "Use a low cost product to attract consumers. Once our organization, via our low cost product, has established a relationship with consumers, our organization will sell additional, higher-margin products and services that enhance the consumer's interaction with the low-cost product or service."

A strategy consists of a well thought out series of tactics to make a marketing plan more effective. Marketing strategies serve as the fundamental underpinning of marketing plans designed to fill market needs and reach marketing objectives. Plans and objectives are generally tested for measurable results.

A marketing strategy often integrates an organization's marketing goals, policies, and action sequences (tactics) into a cohesive whole. Similarly, the various strands of the strategy, which might include advertising, channel marketing, internet marketing, promotion and public relations can be orchestrated. Many companies cascade a strategy throughout an organization, by creating strategy tactics that then become strategy goals for the next level or group. Each one group is expected to take that strategy goal and develop a set of tactics to achieve that goal. This is why it is important to make each strategy goal measurable.

Marketing strategies are dynamic and interactive. They are partially planned and partially unplanned.

7.3.4 Types of Strategies

Marketing strategies may differ depending on the unique situation of the individual business. However there are a number of ways of categorizing some generic strategies. A brief description of the most common categorizing schemes is presented below:

- ◉ **Strategies based on market dominance** - In this scheme, firms are classified based on their market share or dominance of an industry. Typically there are three types of market dominance strategies:

  o   Leader

  o   Challenger

  o   Follower

- ◉ **Porter generic strategies** - strategy on the dimensions of strategic scope and strategic strength. Strategic scope refers to the market penetration while strategic strength refers to the firm's sustainable competitive advantage.

  o   Product differentiation

  o   Market segmentation

- ◉ **Innovation strategies** - This deals with the firm's rate of the new product development and business model innovation. It asks whether the company is on the cutting edge of technology and business innovation. There are three types:

  o   Pioneers

o   Close followers

o   Late followers

- **Growth strategies** - In this scheme we ask the question, "How should the firm grow?". There are a number of different ways of answering that question, but the most common gives four answers:

  o   Horizontal integration

  o   Vertical integration

  o   Diversification

  o   Intensification

A more detailed scheme uses the categories:

- Prospector

- Analyzer

- Defender

- Reactor

- **Marketing warfare strategies** - This scheme draws parallels between marketing strategies and military strategies.

7.4 STRATEGIC MODELS

Marketing participants often employ strategic models and tools to analyze marketing decisions. When beginning a strategic analysis, the 3Cs can be employed to get a broad understanding of the strategic environment. An Ansoff Matrix is also often used to convey an organization's strategic positioning of their marketing mix. The 4Ps can then be utilized to form a marketing plan to pursue a defined strategy.

7.4.1 Marketing in Practice

The Consusmer-Centric Business

There are a many companies especially those in the Consumer Package Goods (CPG) market that adopt the theory of running their business centered around Consumer, Shopper & Retailer needs. Their Marketing departments spend quality time looking for "Growth Opportunities" in their categories by identifying relevant insights (both mindsets and behaviors) on their target Consumers, Shoppers and retail partners. These Growth Opportunities emerge from changes in market trends, segment dynamics changing and also internal brand or operational business challenges. The Marketing team can then prioritize these Growth Opportunities and begin to develop strategies to exploit the opportunities that could include new or adapted products, services as well as changes to the 7Ps.

Real-life marketing primarily revolves around the application of a great deal of common-sense; dealing with a limited number of factors, in an environment of imperfect information and limited resources complicated by uncertainty and tight time scales. Use of classical marketing techniques, in these circumstances, is inevitably partial and uneven.

Thus, for example, many new products will emerge from irrational processes and the rational development process may be used (if at all) to screen out the worst non-runners. The design of the advertising, and the packaging, will be the output of the creative minds employed; which management will then screen, often by 'gut-reaction', to ensure that it is reasonable.

For most of their time, marketing managers use intuition and experience to analyze and handle the complex, and unique, situations being faced; without easy reference to theory. This will often be 'flying by the seat of the pants', or 'gut-reaction'; where the overall strategy, coupled with the knowledge of the customer which has been absorbed almost by a process of osmosis, will determine the quality of the marketing employed. This, almost instinctive management, is what is sometimes called 'coarse marketing'; to distinguish it from the refined, aesthetically pleasing, form favored by the theorists.

7.4.2 Marketing Plan

A marketing plan is a written document that details the necessary actions to achieve one or more marketing objectives. It can be for a product or service, a brand, or a product line. Marketing plans cover between one and five years.

A marketing plan may be part of an overall business plan. Solid marketing strategy is the foundation of a well-written marketing plan. While a marketing plan contains a list of actions, a marketing plan without a sound strategic foundation is of little use.

The Marketing Planning Process

The marketing process model based on the publications of Philip Kotler. It consists of 5 steps, beginning with the market & environment research. After fixing the targets and setting the strategies, they will be realised by the marketing mix in step 4. The last step in the process is the marketing controlling.

In most organizations, "strategic planning" is an annual process, typically covering just the year ahead. Occasionally, a few organizations may look at a practical plan which stretches three or more years ahead.

To be most effective, the plan has to be formalized, usually in written form, as a formal "marketing plan." The essence of the process is that it moves from the general to the specific; from the overall objectives of the organization down to the individual action plan for a part of one marketing program. It is also an interactive process, so that the draft output of each stage is checked to see what impact it has on the earlier stages - and is amended.

Marketing Planning Goals and Objectives

Behind the corporate objectives, which in themselves offer the main context for the marketing plan, will lay the "corporate mission"; which in turn provides the context for these corporate objectives.

This "corporate mission" can be thought of as a definition of what the organization is; of what it does: "Our business is …". This definition should not be too narrow, or it will constrict the development of the organization; a too rigorous concentration on the view that "We are in the business of making meat-scales," as IBM was during the early 1900s, might have limited its subsequent development into other areas. On the other hand, it should not be too wide or it will become meaningless; "We want to make a profit" is not too helpful in developing specific plans.

Abell suggested that the definition should cover three dimensions: "customer groups" to be served, "customer needs" to be served, and "technologies" to be utilized. Thus, the definition of IBM's "corporate mission" in the 1940s might well have been: "We are in the business of handling accounting information [customer need] for the larger US organizations [customer group] by means of punched cards [technology]."

7.5 ENTERPRISE MARKETING MANAGEMENT

Enterprise Marketing Management defines a category of software used by marketing operations to manage their end-to-end internal processes. EMM is subset of Marketing Technologies which consists of a total of 3 key technology types that allow for corporations and customers to participate in a holistic and real-time marketing campaign.

EMM consists of other marketing software categories such as Web Analytics, Campaign Management, Digital Asset Management, Web Content Management, Marketing Resource Management, Marketing Dashboards, Lead Management, Event-driven Marketing, Predictive Modeling and more. The goal of deploying and using EMM is to improve both the efficiency and effectiveness of marketing by increasing operational efficiency, decreasing costs and waste, and standardizing marketing processes for an accurate and predictable time to market. The benefit of using an EMM suite rather than a variety of point solutions is improved collaboration, efficiency and visibility across the entire marketing function, as well as reduced total cost of ownership. Depending on the variable combinations of solutions, EMM can mean several different things to specific brands and industries. Enterprise Marketing Management allows for corporations to put in place a baseline of their operations that will allow them to begin evolution towards a holistic solution that incorporates customer experience, expectation and brand value associated with Marketing Technologies.

7.5.1 Marketing Performance Measurement and Management

Marketing performance measurement and management (MPM) is a term used by marketing professionals to describe the analysis and improvement of the efficiency and effectiveness of marketing. This is accomplished by focus on the alignment of marketing activities, strategies, and metrics with business goals. It involves the creation a metrics framework to monitor marketing performance, and then develop and utilize marketing dashboards to manage marketing performance. This strategy is used by several companies such as IBM, Intel, and Citrix.

Performance management is one of the key processes applied to business operations such as manufacturing, logistics, and product development. The goals of performance management are to achieve key outcomes and objectives to optimize individual, group, or organizational performance. MPM however, is more specific. It focuses on measuring, managing, and analyzing marketing performance to maximize effectiveness and optimize the return of investment (ROI) of marketing. Three elements play a critical role in managing marketing performance—data, analytics, and metrics.

7.5.2 Data and Analytics

One of the core methodologies to measure the effectiveness of marketing is the collection of appropriate data. The gathering of right type of data, and its accuracy, is crucial in measuring the marketing performance. A consonance among the marketing department and the senior management of the organization is important in selecting the appropriate data that have to be collected.

While data collection is relatively simple, a thorough analysis to make sense of collected data is business-critical. By thoroughly analyzing the data, organizations can gather actionable business insights to improve the marketing effectiveness and marketing efficiency. For example, organizations can use the analytics can to drive the marketing return on investment, and make faster and better business decisions.

One of the most common uses of analytics that marketing performance management focuses on optimizing marketing spend by using market mix models. These measure the impact of marketing activities, competitive effects, and market environment on sales of a product to improve marketing effectiveness and drive return on marketing investment. Consumer packaged goods (CPG) industry extensively uses this method, although it is now being adopted in retail, telecom, and financial services industries. Market-mix models use data to create a model that establishes the link between spend in various channels, geographies and so on with incremental sales. The concepts and tools of a marketing mix modeling date back over 30 years. With the increased usage of the Internet, social networking sites, mobile advertising, and text messaging, marketing professionals have revived an interest in market mix models.

7.6 METRICS AND MANAGEMENT

Measurement and metrics enable marketing professionals to justify budgets based on returns and to drive organizational growth and innovation. As a result, marketers use these metrics and performance measurement as way to prove value and demonstrate the contribution of marketing to the organization.

Popular metrics used in analysis include activity-based metrics that involves numerical counting and reporting. For example, tracking downloads, Web site visitors, attendees at various events are types of activity-based metrics. However, they seldom link marketing to business outcomes. Instead, business outcomes such as market share, customer value, and new product adoption offer a better correlation. MPM focuses on measuring the aggregated effectiveness and efficiency of the marketing organization. Some common categories of these specific metrics include marketing's impact on share of preference, rate of customer acquisition, average order value, rate of new product and service adoptions, growth in customer buying frequency, volume and share of business, net advocacy and loyalty, rate of growth compared to competition and the market, margin, and customer engagement. In addition, MPM is used to measure the monitoring of operational efficiency and external performance.

Operations performance metrics is a term used when organizations manage marketing functions as a business. Organizations committed to implementing MPM may create positions such as marketing operations director and marketing finance director. Program-to-people ratios, awareness-to-demand ratios, the cost vs lead, the cost vs sale, and conversation rates are the typical data collected and analyzed. Operational performance metrics, however, primarily provide the organization with a way to rationalize marketing investments, but do not correlate marketing to business strategy and business performance.

MPM tightly focuses on these operating measures to help marketers view how efficiently resources of the organization such as people, facilities, and capital are used. External performance measures aligned with business outcomes assess things such as the value an organization provides to customers or the performance of an organization relative to its competitors.

By using a top-down approach, marketers develop metrics and specific performance targets known as key performance indicators (KPI). First business decisions are made to define the scope. To create metrics and KPIs, marketeers involved in MPM try to first brainstorm on the business outcome that they are trying to impact. This is followed by asking the opposite questions that need to be answered to determine if the questions have an impact on this outcome, and the necessary supporting data required to answer these questions. After determining what data is needed, marketeers need to search for this data, and determine the decisions and actions that must to be enforced as a result of this data mining.

7.6.1 Management and Skills

A Forrester Research and Heidrick and Struggles study titled The Evolved CMO, found that 20% of the 115 chief and senior marketing professionals needed a further understanding in marketing measurement, customer relationship management (CRM), and customer data analytics. Marketing professionals must be able to tap into customer information that enables them to provide a strategic guidance the organization requirements. This allows them extend business into emerging markets and bring innovative products to market.

Professionals skilled in measurement, analytics, and data-management can serve as catalysts of growth. They can anticipate customer requirement and develop marketing capabilities of their organization. These marketing professionals can then assess how to measure the impact of marketing on the business and translate the metrics to simple charts, data, and figures that the top management of the organization can understand.

A United States-based study conducted by management consultants Bersin and Associates state that "today more than 40% of US corporations believe that 'driving a performance-based culture' is one of their top three talent strategies." Cultures that foster an environment of teamwork and employee development, and resource empowerment achieve higher quality outcomes. Similar to other optimization processes, MPM requires a culture of accountability within the organization. Without a disciplined approach, success of implementation of MPM to drive productivity would not yield tangible results.

7.6.2 Measurement of Progress

The final stage of any marketing planning process is to establish targets (or standards) so that progress can be monitored. Accordingly, it is important to put both quantities and timescales into the marketing objectives (for example, to capture 20 percent by value of the market within two years) and into the corresponding strategies.

Changes in the environment mean that the forecasts often have to be changed. Along with these, the related plans may well also need to be changed. Continuous monitoring of performance, against predetermined targets, represents a most important aspect of this. However, perhaps even more important is the enforced discipline of a regular formal review. Again, as with forecasts, in many cases the best (most realistic) planning cycle will revolve around a quarterly review. Best of all, at least in terms of the quantifiable aspects of the plans, if not the wealth of backing detail, is probably a quarterly rolling review - planning one full year ahead each new quarter. Of course, this does absorb more planning resource; but it also ensures that the plans embody the latest information, and - with attention focused on them so regularly - forces both the plans and their implementation to be realistic.

Plans only have validity if they are actually used to control the progress of a company: their success lies in their implementation, not in the writing'.

7.6.3 Performance Analysis

The most important elements of marketing performance, which are normally tracked, are:

Sales analysis : Most organizations track their sales results; or, in non-profit organizations for example, the number of clients. The more sophisticated track them in terms of "sales variance" - the deviation from the target figures - which allows a more immediate picture of deviations to become evident.. `Micro- analysis`, which is a nicely pseudo-scientific term for the normal management process of investigating detailed problems, then investigates the individual elements (individual products, sales territories, customers and so on) which are failing to meet targets.

Market share analysis : Few organizations track market share though it is often an important metric. Though absolute sales might grow in an expanding market, a firm's share of the market can decrease which bodes ill for future sales when the market starts to drop. Where such market share is tracked, there may be a number of aspects which will be followed:

- Overall market share
- Segment share - that in the specific, targeted segment
- Relative share -in relation to the market leaders
- Annual fluctuation rate of market share

Expense analysis : The key ratio to watch in this area is usually the 'marketing expense to sales ratio'; although this may be broken down into other elements (advertising to sales, sales administration to sales, and so on).

Financial analysis : The "bottom line" of marketing activities should at least in theory, be the net profit (for all except non-profit organizations, where the comparable emphasis may be on remaining within budgeted costs). There are a number of separate performance figures and key ratios which need to be tracked:

- Gross contribution<>net profit
- Gross profit<>return on investment
- Net contribution<>profit on sales

There can be considerable benefit in comparing these figures with those achieved by other organizations (especially those in the same industry); using, for instance, the figures which can be obtained (in the UK) from 'The Centre for Interfirm Comparison'. The most sophisticated use of this approach, however, is typically by those making use of PIMS (Profit Impact of Management Strategies), initiated by the General Electric Company and then developed by Harvard Business School, but now run by the Strategic Planning Institute.

The above performance analyses concentrate on the quantitative measures which are directly related to short-term performance. But there are a number of indirect measures, essentially tracking customer attitudes, which can also indicate the organization's performance in terms of its longer-term marketing strengths and may accordingly be even more important indicators. Some useful measures are:

⊙ **Market research** - including customer panels (which are used to track changes over time)

⊙ **Lost business** - the orders which were lost because, for example, the stock was not available or the product did not meet the customer's exact requirements

⊙ **Customer complaints** - how many customers complain about the products or services, or the organization itself, and about what

Use of marketing plans : A formal, written marketing plan is essential; in that it provides an unambiguous reference point for activities throughout the planning period. However, perhaps the most important benefit of these plans is the planning process itself. This typically offers a unique opportunity, a forum, for information-rich and productively focused discussions between the various managers involved. The plan, together with the associated discussions, then provides an agreed context for their subsequent management activities, even for those not described in the plan itself.

7.7 BUDGETS AS MANAGERIAL TOOLS

The classic quantification of a marketing plan appears in the form of budgets. Because these are so rigorously quantified, they are particularly important. They should, thus, represent an unequivocal projection of actions and expected results. What is more, they should be capable of being monitored accurately; and, indeed, performance against budget is the main (regular) management review process.

The purpose of a marketing budget is, thus, to pull together all the revenues and costs involved in marketing into one comprehensive document. It is a managerial tool that balances what is needed to be spent against what can be afforded, and helps make choices about priorities. It is then used in monitoring performance in practice.

The marketing budget is usually the most powerful tool by which you think through the relationship between desired results and available means. Its starting point should be the marketing strategies and plans, which have already been formulated in the marketing plan itself; although, in practice, the two will run in parallel and will interact. At the very least, the rigorous, highly quantified, budgets may cause a rethink of some of the more optimistic elements of the plans.

7.7.1 Value of Marketing

Value of a product within the context of marketing means the relationship between the consumer's expectations of product quality to the actual amount paid for it. It is often expressed as the equation :

Value = Benefits/Price

or alternatively:

Value = Quality received / Expectations

There are parallels between cultural expectations and consumer expectations. Thus pizza in Japan might be topped with tuna rather than pepperoni, as pizza might be in the US; the value in the marketplace varies from place to place as well as from market to market.

For a firm to deliver value to its customers, they must consider what is known as the "total market offering." This includes the reputation of the organization, staff representation, product benefits, and technological characteristics as compared to competitors' market offerings and prices. Value can thus be defined as the relationship of a firm's market offerings to those of its competitors.

Value in marketing can be defined by both qualitative and quantitative measures. On the qualitative side, value is the perceived gain composed of individual's emotional, mental and physical condition plus various social, economic, cultural and environmental factors. On the quantitative side, value is the actual gain measured in terms of financial numbers, percentages, and dollars.

For an individual to deliver value, one has to grow his / her knowledge and skill sets to showcase benefits delivered in a transaction (e.g., getting paid for a job).

For an organization to deliver value, it has to improve its value : cost ratio. When an organization delivers high value at high price, the perceived value may be low. When it delivers high value at low price, the perceived value may be high. The key to deliver high perceived value is attaching value to each of the individuals or organizations — making them believe that what you are offering is beyond expectation — helping them to solve a problem, offering a solution, giving results, and making them happy.

Value changes based on time, place and people in relation to changing environmental factors. It is a creative energy exchange between people and organizations in our marketplace.

7.8 MARKETING CHANNELS

Meaning: Farmers producing agricultural produce are scattered in remote villages while consumers are in semi-urban and urban areas. This produce has to reach consumers for its final use and consumption. There are different agencies and functionaries through which this produce passes and reaches the consumer. A market channel or channel of distribution is therefore defined as a path traced in the direct or indirect transfer of title of a product as it moves from a producer to an ultimate consumer or industrial user. Thus, a channel of distribution of a product is the route taken by the ownership of goods as they move from the producer to the consumer or industrial user.

Factors affecting channels: There are several channels of distribution depending upon type of produce or commodity. Each commodity group has slightly different channel. The factors are :

1. Perishable nature of produce.e.g. fruits, vegetables, flowers, milk, meat, etc.

2. Bulk and weight-cotton, fodders are bulky but light in weight.

3. Storage facilities.

4. Weak or strong marketing agency.

5. Distance between producer and consumer. Whether local market or distant market.

7.8.1 Types of Market Channels

Some of the typical marketing channels for different product groups are given below:

Channels of Rice

1. Producer-miller->consumer (village sale)

2. Producer-miller->retailer-consumer (local sale)

3. Producer-wholesaler->miller-retailer-consumer

4. Producer-miller-cum-wholesaler-retailer-consumer

5. Producer-village merchant-miller-retailer-consumer

6. Producer-govt. procurement-miller-retailer-consumer

Channels of Other Foodgrains

1. Producer - consumer (village sale)

2. Producer-village merchant-consumer (local sale)

3. Producer-wholesaler-cum-commission agent retailer-consumer

4. Producer-primary wholesaler-secondary wholesaler- retailer- Consumer

5. Producer-Primary wholesaler-miller-consumer (Bakers).

6. Producer->govt.procurement-retailer-consumer.

7. Producer-government-miller-retailer-consumer.

Channels of Cotton

1. Producer-village merchant-wholesaler or ginning factory- wholesaler in lint-textile mill (consumer)

2. Producer-Primary wholesaler-ginning factory-secondary wholesaler-consumer (Textile mill)

3. Producer- Trader- ginning factory- wholesaler in lint- consumer (Textile mill)

4. Producer-govt. agency-ginning factory-consumer (Textile mill).

5. Producer-Trader-ginning factory-wholesaler-retailer- consumer (non-textile use).

Channels of Vegetables

1. Producers-consumer (village sale)
2. Producer-retailer-consumer (local sale)
3. Producer-Trader-commission agent-retailer-consumer.
4. Producer-commission agent-retailer-consumer
5. Producer-primary wholesaler-secondary wholesaler- retailer- consumer (distant market).

Channels of Fruits

1. Producer-consumer (village sale)
2. Producer-Trader-consumer (local sale)
3. Producer-pre-harvest contractor-retailer-consumer
4. Producer-commission agent-retailer-consumer.
5. Producer-pre-harvest contractor-commission agent- retailer-consumer
6. Producer-commission agent-secondary wholesaler- retailer-consumer (distant market). These channels have great influence on marketing costs such as transport, commission charges, etc. and market margins received by the intermediaries such as trader, commission agent, wholesaler and retailer. Finally this decides the price to be paid by the consumer and share of it received by the farmer producer. That channel is considered as good or efficient which makes the produce available to the consumer at the cheapest price also ensures the highest share to the producer.

7.9 AGRICULTURAL MARKETING DEVELOPMENT

Efforts to develop agricultural marketing have, particularly in developing countries, tended to concentrate on a number of areas, specifically infrastructure development; information provision; training of farmers and traders in marketing and post-harvest issues; and support the development of an appropriate policy environment. In the past, efforts were made to develop government-run marketing bodies but these have tended to become less prominent over the years.

7.9.1 Agricultural Market Infrastructure

Efficient marketing infrastructure such as wholesale, retail and assembly markets and storage facilities is essential for cost-effective marketing, to minimize post-harvest losses and to reduce health risks. Markets play an important role in rural development, income generation, food security, and developing rural-market linkages. Experience shows that planners need to be aware of how to design markets that meet a community's social and economic needs and how to choose a suitable site for a new market. In many cases sites are chosen that are inappropriate and result in under-use or even no use of the infrastructure constructed. It is also not sufficient just to build a market: attention needs to be paid to how that market will be managed, operated and maintained. Most market improvements that have been only aimed at infrastructure upgrading and have not guaranteed maintenance and management have failed within a few years

Rural assembly markets are located in production areas and primarily serve as places where farmers can meet with traders to sell their products. These may be occasional (perhaps weekly) markets, such as haat bazaars in India and Nepal, or permanent. Terminal wholesale markets are located in major metropolitan areas, where produce is finally channelled to consumers through trade between wholesalers and retailers, caterers, etc. The characteristics of wholesale markets have changed considerably as retailing changes in response to urban growth, the increasing role of supermarkets and increased consumer spending capacity. These changes may require responses in the way in which traditional wholesale markets are organized and managed.

Retail marketing systems in western countries have broadly evolved from traditional street markets through to the modern hypermarket or out-of-town shopping center. In developing countries, there remains scope to improve agricultural marketing by constructing new retail markets, despite the growth of supermarkets, although municipalities often view markets primarily as sources of revenue rather than infrastructure requiring development. Effective regulation of markets is essential. Inside a market, both hygiene rules and revenue collection activities have to be enforced. Of equal importance, however, is the maintenance of order outside the market. Licensed traders in a market will not be willing to cooperate in raising standards if they face competition from unlicensed operators outside who do not pay any of the costs involved in providing a proper service

7.10 MARKET INFORMATION

Efficient market information can be shown to have positive benefits for farmers and traders. Up-to-date information on prices and other market factors enables farmers to negotiate with traders and also facilitates spatial distribution of products from rural areas to towns and between markets. Most governments in developing countries have tried to provide market information services to farmers, but these have tended to experience problems of sustainability. Moreover, even when they function, the service provided is often insufficient to allow commercial decisions to be made because of time lags between data collection and dissemination. Modern communications technologies open up the possibility for market information services to improve information delivery through SMS on cell phones and the rapid growth of FM radio stations in many developing countries offers the possibility of more localised information services. In the longer run, the internet may become an effective way of delivering information to farmers. However, problems associated with the cost and accuracy of data collection still remain to be addressed. Even when they have access to market information, farmers often require assistance in interpreting that information. For example, the market price quoted on the radio may refer to a wholesale selling price and farmers may have difficulty in translating this into a realistic price at their local assembly market. Various attempts have been made in developing countries to introduce commercial market information services but these have largely been targeted at traders, commercial farmers or exporters. It is not easy to see how small, poor farmers can generate sufficient income for a commercial service to be profitable although in India a new service introduced by Thomson Reuters was reportedly used by over 100,000 farmers in its first year of operation. Esoko in West Africa attempts to subsidize the cost of such services to farmers by charging access to a more advanced feature set of mobile-based tools to businesses.

7.10.1 Marketing Training

Farmers frequently consider marketing as being their major problem. However, while they are able to identify such problems as poor prices, lack of transport and high post-harvest losses, they are often poorly equipped to identify potential solutions. Successful marketing requires learning new skills, new techniques and new ways of obtaining information. Extension officers working with ministries of agriculture or NGOs are often well-trained in horticultural production techniques but usually lack knowledge of marketing or post-harvest handling.

7.10.2 Agricultural Marketing Support

Most governments have at some stage made efforts to promote agricultural marketing improvements. In the United States the Agricultural Marketing Service (AMS) is a division of USDA and has programs that provide testing, support standardization and grading and offer market news services. AMS oversees marketing agreements and orders research and promotion programs. It also purchases commodities for federal food programs. USDA also provides support to agricultural marketing work at various universities. In the United Kingdom, support for marketing of some commodities was provided before and after the Second World War by boards such as the Milk Marketing Board and the Egg Marketing Board. These boards were closed down in the 1970s. As a colonial power, Britain established marketing boards in many countries, particularly in Africa. Some continue to exist although many were closed at the time of the introduction of structural adjustment measures in the 1990s.

Several developing countries have established government-sponsored marketing or agribusiness units. South Africa, for example, started the National Agricultural Marketing Council (NAMC) as a response to the deregulation of the agriculture industry and closure of marketing boards in the country. India has the long-established National Institute of Agricultural Marketing. These are primarily research and policy organizations, but other agencies provide facilitating services for marketing channels, such as the provision of infrastructure, market information and documentation support. Examples from the

Caribbean include the National Agricultural Marketing Development Corporation (NAMDEVCO) in Trinidad and Tobago and the New Guyana Marketing Corporation in Guyana.

7.11 AGRICULTURAL VALUE CHAIN

The agricultural value chain concept has been used since the beginning of the millennium, primarily by those working in agricultural development in developing countries. Although there is no universally accepted definition of the term, it normally refers to the whole range of goods and services necessary for an agricultural product to move from the farm to the final customer or consumer.

7.11.1 Agricultural Value Chain Finance

Agricultural value chain finance is concerned with the flows of funds to and within a value chain to meet the needs of chain actors for finance, to secure sales, to buy inputs or produce, or to improve efficiency. Examining the potential for value chain finance involves a holistic approach to analyze the chain, those working in it, and their inter-linkages. These linkages allow financing to flow through the chain. For example, inputs can be provided to farmers and the cost can be repaid directly when the product is delivered, without need for farmers taking a loan from a bank or similar institution. This is common under contract farming arrangements. Types of value chain finance include product financing through trader and input supplier credit or credit supplied by a marketing company or a lead firm. Other trade finance instruments include receivables financing where the bank advances funds against an assignment of future receivables from the buyer, and factoring in which a business sells its accounts receivable at a discount. Also falling under value chain finance are asset collateralization, such as on the basis of warehouse receipts, and risk mitigation, such as forward contracting, futures and insurance

7.11.2 Links Between Agriculture and the Food Industry

The link between agriculture and food continually evolves. In primitive societies, the farmer and consumer were either the same family or close neighbours who bartered their products and services as we see in figure 1.1, but as societies develop other linkages are added. Commodity traders, processors, manufacturers who convert produce into food items and retailers, among others, are interposed between the producer and consumer. A more recently introduced link into the chain is the scientist. Scientists as breeders, plant biologists, nutritionists and chemists have made an immeasurable contribution to the development of agricultural production and food manufacture over the past 50 years. It would appear that we have passed through the age of machines in agriculture, and the age of chemicals, on to the age of biotechnology in agriculture. Biotechnology has great potential for the developing countries since it is likely to be less capital intensive and more research and know-how intensive. Thus its benefits can flow faster into the poorer countries who do not have the capital. Therefore its impact could be faster, more widespread and more significant.

As the link between food and agriculture continues to evolve, we see the emergence of an agribusiness i.e. where agriculture and food become a continuum. Multinational companies like Cargill, Brooke Bond Liebig, and Del Monte are examples of vertically integrated organisations with links all the way through from agricultural production to retailing. There is a line of argument which says that it makes sense that those who are closest should the consumer should assess his/her needs and interpret them back to the primary producer.

As disposable incomes increase, the food industry will increase the quality and diversity of the products it produces. Food manufacturers will have particular expectations of agriculture as a supplier of their raw materials, including:

Quality: To build a profitable business, food manufacturers seek to establish a preference for their products by differentiating those products in some way which is meaningful to consumers. Then, in order to enable consumers to recognise the differentiated product, manufacturers brand that product. Manufacturers can then work on building consumer loyalty to these brands. Brand loyalty is normally only established by delivering high quality consistently. As disposable incomes rise, the market tends to develop more sophisticated needs and the quality of the raw material becomes even more critical. Where agriculture is seeking to serve a food industry, that itself is seeking to meet these more sophisticated needs and wants, it can expect to face increasing emphasis on quality. Equally well, agriculture can expect to share in the better return for innovative improvements in quality.

Cost: Next to quality will come cost. With an increased capability to search the world for raw materials, the food industry is able to find the lowest cost source for any given level of quality. For the food manufacturer, the country in which he/she manufactures, or markets, need no longer be the source of agricultural produce. Improved transportation and communications mean that the world is becoming his/her source of supply. This is a significant change in the competitive environment of agriculture which the farming community has to realise, because they have, hitherto, been largely concooned in their respective domestic markets.

Non-seasonality: Agricultural products were traditionally seasonal in their production and supply. Modern technology and husbandry practices mean that food manufacturers need not have their production schedules dictated by the seasons. Indeed, the capital intensive food industry cannot afford to incur the high costs of under utilising its capacity. This means that farmers will have to complete in terms of reducing seasonality or fitting into a pattern of social competitiveness.

Reliability: A manufacturer who has invested heavily in building up his brand will be very keen to get reliable supplies in terms of quality, timing and cost. Producers of agricultural produce will be increasingly judged on their reliability in all of these respects.

Processing: Ease of processing will become an increasingly important expectation of the food industry. Like all industries, reductions in the costs of capital equipment, wages and inventories are important objectives. For example, farmers who can deliver on the 'just-in-time' principle will contribute towards reducing a manufacturer's working capital and space requirements. Farmers who can do part of the secondary processing and/or performing functions such as the post harvest treatment of the crop or transporting will be adding another advantage. Crops that are specially bred or designed to facilitate processing (e.g. seedless fruits, featherless chickens, coffee beans without caffeine, low cholesterol meats) are another type of advantage that the food industry could expect from agriculture. In short, the competitive advantage will rest with those able to add most value and can differentiate what they are offering from that of other suppliers.

Product differentiation: In competitive brand marketing, the food industry has to innovate continuously to create new products that are different from and superior to existing ones of their own or competitors. The scope of innovation has traditionally been at the processing stage. Whilst this will continue to be an important area for innovation, manufacturers will increasingly tend to look for innovative changes in the agricultural produce itself. This may be in terms of novel tastes, improved texture, more attractive shapes, etc.

Health aspects: We have already said that in the more sophisticated food markets, healthy eating can become a priority among consumers. Therefore, farmers will have to consider the health connotations of what they choose to grow. There are two aspects of health to be taken into account. First, consumers may be interested in the food itself i.e. low fat, low/no sugar or low/no salt. It would be a mistake to think that health issues are confined to the more sophisticated food markets or to the wealthier segments of the community. Nutrition is important in all segments of the market. Even where the poor receive adequate amounts of food to fend off starvation, they are often malnourished. Thus farmers have to be concerned about the nutritional value of the produce they grow. Second, the consumer may be more, or equally, concerned about the food production methods i.e. the avoidance of chemicals like herbicides, pesticides etc. This may mean a change to the farmer's husbandry practices with implications for the costs of production. The consumer and the food industry will expect the farmer to produce without potentially dangerous chemicals, but at no extra cost to them. This will be another challenge for agriculture.

7.12 CO-OPERATIVES IN THE AGRICULTURE AND FOOD SECTORS

The co-operative enterprise has its origins in the 19th century and has become one of the most ubiquitous examples forms of business/economic enterprise. Co-operatives exist in all countries of the world and operate under diverse political systems: from communism to capitalism. The majority of these co-operatives are, through their national apex organisations, ultimately in membership of the International Co-operative Alliance (ICA), the representative world body of co-operatives of all types.

The motivation to form co-operatives has three particular aspects:

- ◉ The need for protection against exploitation by economic forces too strong for the individual to withstand alone

- ◉ The impulse for self-improvement by making the best use of often scarce resources

- ◉ The concern to secure the best possible return from whatever from of economic activity within which the individual engages whether as a producer, intermediary or consumer.

It is the belief that each of these aspirations can most advantageously be pursued and secured in concert with like-minded people that provides the stimulus to co-operative action. The underpinning principles with are those of self-help, voluntary participation, equity, democracy, and a common bond of common need and purpose. The cohesion of the group is maintained by ensuring that individual members cannot secure power or gain advantages at the expense of the others. Co-operatives reward participation in the co-operative venture rather than rewarding capital Self-interest is a primary motivator in co-operative enterprises, with economic gain being the primary objective. In these respects, co-operatives differ little from capitalistic enterprises; self-interest is simply pursued in a different way from the capitalist enterprise. Thus, the rate of interest paid on share capital is fixed and limited, and not subject to variation according to the amount of profit made. Secondly the use and distribution of surplus is restricted to one or more of the following purposes:

- ◉ Allocation to reserves, where it becomes collectively-owned capital and is thereafter non-distributable

- ◉ For use on, or donation to, common-good, community project

- ◉ Distribution to members in proportion to the trade each member has done with the co-operative. In other words, the distribution is made not in relation to capital held, but by declaring a bonus or dividend per cash unit of trade done

Why agricultural diversification is vital for our rural transformation:

Diversification of agriculture refers to the shift from the regional dominance of one crop to regional production of a number of crops to meet ever increasing demand for both cash crop and food production. It takes into account the economic returns from different value-added crops with complementary marketing opportunities.

A diversified portfolio of products ensures that farmers don't suffer complete ruin when the weather is unpredictable. It manages price risk, on the assumption that not all products will suffer low prices at the same time. Unfortunately, most farmers often do the opposite of diversification by planting products that have a high price in one year, only to see the price collapse in the next.

Diversification in agriculture is key in achieving food security, improved human nutrition and increase in rural employment. Without diversification, farmers who are dependent on exports run a number of risks. A classic example is the Caribbean banana industry, which collapsed as a result of the removal of quota protection on EU markets, necessitating diversification by the region's farmers.

Farmers in several countries, including, India, Kenya, Sri Lanka, etc., have already initiated diversification as a response to climate change. For instance, government policy in Kenya to promote crop diversification has included the removal of subsidies for some crops, encouraging land-use zoning and introducing differential land tax systems.

7.12.1 The Case of Agricultural Diversification in India

The decline in the share of agriculture in India's GDP has been rapid in the post-liberalisation era; despite the fact that the growth of agricultural income during the 1990s has been marginally better than the corresponding rate of growth in the 1980s. Growth in agriculture stagnated towards the end of the 1990s and has decelerated thereafter. In this context, the composition of income from agriculture and allied sectors of the economy has been suited.

Agricultural diversification has been achieved as a larger area of land is now utilised for the cultivation of high value crops. With trade liberalisation, the relative prices of exportable commodities have increased rapidly and those of importable commodities have comparatively decreased. In the short run, a continuous increase in the relative price of a commodity enhances its production more often by substituting it for importable commodities without any statistical effect on the cropped area. As a result, the share of exportable commodities has increased in the total value of agricultural output.

Considering the multi-dimensional importance of agriculture diversification, it is important to understand the drivers of agricultural diversification in the country. Changes in the relative prices of corps have influenced the crop enterprise mix immensely. Price is basically a reflection of the demand and supply situation. In a closed economy, the prices that farmers receive alternately, farm harvest price is influenced by the minimum support price (MSP) and the MSP has been influencing acreage under crops. But in several growing economies, MSPs for agricultural commodities have been increased in recent past. Apart from contributing to food price inflation, this increases the spread between prices paid to producers and subsidised prices charged to consumers, fueling the fiscal burden. Since MSPs need not always extend to all agricultural commodities and public procurement need not cover all commodities either, this creates perverse price signals and distorts resource allocation.

Triggered by food price increases, there have been interventions on the consumption side, including price controls, consumption subsidies, food aid, food for work arrangements, cash transfers and the elimination of taxes on consumption across a range of countries. Are these fiscally sustainable? Do they lead to additional distortions? Do they lead to supply-side adjustments or are they knee-jerk reactions? But in India's case, agricultural diversification has not affected much through the price rise or policy response directed on that -so the case of agricultural diversification remains positive in India.

The size and the quality of land has always been an important factor in agricultural production relations. The average size of operational holding is often considered an important determinant of crop diversification. These variables are supposed to have a negative effect on diversification indices. Many populous states of India such as Bihar and UP are witnessing the growth of small size of farmland, which is affecting productivity and income involved with this. After an extent, the role of technology in improving productivity while alternately reducing the unit cost of production and conserving natural resources cannot be overemphasised. Thus, land management is the need of hour to sustain the welcome pattern of agricultural diversification.

The Declaration on Agricultural Diversification was launched and signed in Paris on 7 December 2015 by world renowned research leaders during the recent meeting of the United Nations Framework Convention on Climate Change (UNFCCC) COP21. The Declaration, which aims to build agricultural resilience and food security for future climates, will underpin the Global Action Plan for Agricultural Diversification (GAPAD) to be developed during 2016.

The Declaration was signed in the presence of Former Malaysian Prime Minister, Tun Abdullah Ahmad Badawi. Founding signatories included Dr Trevor Nicholls of the Centre for Agriculture and Bioscience (CABI), Dr David Molden of International Centre for Integrated Mountain Development (ICIMOD), Dr Jose Joachim Campos of Centro Agronomico de Investigacion Ensenanza (CATIE), Prof. Meine Van Noorwijk of the World Agroforestry Centre (ICRAF) and representatives from the Association of International Research and Development Centers for Agriculture (AIRCA), Food and Agriculture Organisation of the United Nations (FAO) and Global Alliance for Climate-Smart Agriculture (GACSA).

7.12.2 Marketing Systems: Functions, Agents, Enterprises and Channels

All marketing systems have evolved within the constraints and conditions placed upon them by the production sector and by the nature of the goods being marketed. The type of product, the number, size and density of producers, the infrastructure and the policy and institutional environments all determine the type of marketing system and the effectiveness with which it operates.

A marketing system is comprised of a number of elements: the particular products (e.g. butter only, or butter and raw milk) and their characteristics being transferred from producer to consumer; the characteristics of participants (e.g. the producer, the trader, the consumer); the functions or roles that each participant performs in the market; and the locations, stages, timetable and physical infrastructures involved.

When we talk of describing, quantifying or analysing a particular marketing system, there is an implicit assumption that we can distinguish the elements of that system from other economic activities. Analyses of marketing systems usually include a quantification of the flows and of the value added, costs and profit margins at each stage in the system.

7.13 AGRICULTURAL AND FOOD MARKETING

As individuals within a society become more specialised in their economic activities, they come to rely upon others to supply at least some of the products and services which they need. Thus begins a process of exchange between buyers and sellers. For a while buyers and sellers remain in immediate contact and each party is able to determine what the other needs and values and, therefore, will be willing to exchange. As the economy develops the number and types of exchanges expand, there is a concomitant need for increasingly specialised marketing services such as physical distribution, storage, grading, market information gathering and so. The number of participants also increases with many of the specialised services being provided by intermediaries between the seller and ultimate buyer. Few buyers and sellers are in direct contact with one another and communication between them is channelled through a complex marketing system. This introductory chapter is devoted to exploring the nature of marketing and marketing systems.

7.13.1 The Importance of Agricultural and Food Marketing to Developing Countries

In many countries, and virtually every less developed country (LDC), agriculture is the biggest single industry. Agriculture typically employs over fifty percent of the labour force in LDCs with industry and commerce dependent upon it as a source of raw materials and as a market for manufactured goods. Hence many argue that the development of agriculture and the marketing systems which impinge upon it are at the heart of the economic growth process in LDCs. This has clear implications for agricultural production and the marketing systems that direct that production and distribute the output to the points of its consumption. Subsistence farming is likely to diminish in importance as farmers respond to the increased opportunities that development and urbanisation create; farms are likely to decrease in number whilst increasing in size; and agriculture will probably become less labour intensive and more capital intensive.

So far this discussion has been set in the context of commercial marketing but social marketing should also be acknowledged. Social marketing identifies human needs in non-competitive economies and/or sectors of society and defines the means of delivering products and services to meet these needs. The marketing mix of social marketing strategies is evaluated using quite different criteria from those employed in assessing purely commercial marketing strategies. Criteria such as the percentage of the target population reached with the technology, products, processes or services, quantities produced and distributed and uptake of the product, service or technology are more often employed. Benefits are measured in terms of development goals, such as improved nutritional status or increased rural incomes. The use of economic criteria is usually limited to the latter and to selecting the least-cost strategy to achieve a quantitative goal. However, the criteria used to evaluate commercial marketing strategies should not automatically be eliminated, because these improve the efficiency of some aspects of social marketing strategy without preventing the attainment of social objectives.

7.13.2 The Marketing Concept and Marketing Systems

Marketing is not simply an extension of the production process but its only purpose as Adam Smith emphasised when, in his text The Wealth of Nations (1776), he said that:

"Consumption is the sole end purpose of all production: and the interest of the producer ought to be attended to only so far as it may be necessary for promoting that of the consumer."

Dixie relates what he describes as a definition of marketing which is:

The marketing concept must be adopted throughout not only the entire organisation, but the entire marketing system. A system is a complex of interrelated component parts or sub-systems which have a defined common goal. Thus, an agricultural and food marketing system comprises all of the functions, and agencies who perform those activities, that are necessary in order to profitably exploit opportunities in the marketplace. Each of the components, or sub-systems, are independent of one another but a change in any one of them impacts on the others as well as upon the system as a whole. There is a danger that the marketing concept will be adopted by some parts of the system but not others. Thus, for example, a food manufacturer may be trying hard to implement the marketing concept and offer products that meet the precise needs of a target market. If, however, the manufacturer has to rely upon a farming community that is still very much production oriented, for raw material supplies, then the overall marketing objectives may be frustrated. In the same way, if only some functions are performed according to the marketing concept then the system as a whole may not achieve a market orientation. For instance, the marketing department may set out to serve the market for a high quality fruits and vegetables, for which it can obtain premium prices, but if transportation is performed using the same open-topped bulk carrying wagons used to ship grain and other aggregates then it is unlikely that the enterprise will deliver the product in the right condition for the target market.

7.14 MARKETING FUNCTIONS

A little earlier it was said that a marketing system has two distinct dimensions. One of those dimensions is the institutions, organisations and enterprises which participate in a market and the second is the functions that those participants perform. Kohls and Uhl have classified the functions involved in agricultural and food marketing processes as under three sets of functions of a marketing system

Each of these functions add value to the product and they require inputs, so they incur costs. As long as the value added to the product is positive, most firms or entrepreneurs will find it profitable to compete to supply the service.

7.14.1 Exchange Functions

Buying: The marketing concept holds that the needs of the customer are of paramount importance. A producer can be said to have adopted a market orientation when production is purposely planned to meet specific demands or market opportunities. Thus a contract farmer who wishes to meet the needs of a food processor manufacturing sorghum-based malted drinks will only purchase improved sorghum seed. He/she will avoid any inputs likely to adversely affect the storage and/or processing properties of the sorghum and will continually seek new and better inputs which will add further value to his/her product in the eyes of the customer. In making his/her buying decisions his underlaying consideration will be the effect upon the attractiveness of his/her output to the markets he/she is seeking to serve.

The buyer's motive is the opportunity to maintain or even increase profits and not necessarily to provide, for example, the best quality. Improving quality inevitably increases the associated costs. In some cases the market is insensitive to improvements in quality, beyond some threshold level, does not earn a premium price. Under such circumstances, the grower who perseveres and produces a 'better product', is not market oriented since he/she is ignoring the real needs of the consumer. The most successful agribusiness is the one which yields the largest difference between prices obtained and costs incurred.

Selling: Of the nine functions listed, this is probably the one which people find least difficulty in associating with marketing. Indeed to many the terms marketing and selling are synonymous. Kotler suggests that:

"Most firms practice the selling concept when they have over capacity. Their immediate aim is to sell what they can make rather than to make what they can sell."

There is no denying that 'high pressure selling' is practiced, where the interests of the consumer are far from foremost in the mind of the seller. This is not marketing. Enterprises adopt the marketing philosophy as a result of becoming aware that their own long term objectives can only be realised by consistently providing customer satisfaction. Whereas selling might create a consumer, marketing is about creating a customer. The difference is that marketing is about establishing and maintaining long term relationships with customers.

Selling is part of marketing in the same way that promotion, advertising and merchandising are components, or sub-components of the marketing mix. These all directed towards persuasion and are collectively known as marketing communications; one of the four elements of the marketing mix.

7.14.2 Physical Functions

Storage: An inherent characteristic of agricultural production is that it is seasonal whilst demand is generally continous throughout the year. Hence the need for storage to allow a smooth, and as far as possible, uninterrupted flow of product into the market. Because he is dealing with a biological product the grower does not enjoy the same flexibility as his manufacturing counterpart in being able to adjust the timing of supply to match demand. It would be an exaggeration to suggest that a manufacturer can turn production on and off to meet demand - they too have their constraints- but they have more alternatives than does the agricultural producer. A manufacturer can, for example, work overtime, sub-contract work, and over a longer time horizon, the manufacturer can increase or decrease productive capacity to match the strength of demand.

In agriculture, and especially in LDCs, supply often exceeds demand in the immediate post-harvest period. The glut reduces producer prices and wastage rates can be extremely high. For much of the reminder of the period before the next harvest, the product can be in short supply with traders and consumers having to pay premium prices to secure whatever scarce supplies are to be had. The storage function is one of balancing supply and demand.

Both growers and consumers gain from a marketing system that can make produce available when it is needed. A farmer, merchant, co-operative, marketing board or retailer who stores a product provides a service. That service costs money and there are risks in the form of wastage and slumps in market demand, prices, so the provider of storage is entitled to a reward in the form of profit.

Transportation: The transport function is chiefly one of making the product available where it is needed, without adding unreasonably to the overall cost of the produce. Adequate performance of this function requires consideration of alternative routes and types of transportation, with a view to achieving timeliness, maintaining produce quality and minimising shipping costs.

Of course, processing is not the only way of adding value to a product. Storing products until such times as they are needed adds utility and therefore adds value. Similarly, transporting commodities to purchasing points convenient to the consumer adds value. In short, any action which increases the utility of the good or service to prospective buyers also adds value to that product or service.

7.14.3 Facilitating Functions

The facilitating functions include product standardisation, financing, risk bearing and market intelligence. Facilitating functions are those activities which enable the exchange process to take place. Marketing, in simple terms, is the act of supplying products to someone in exchange for something perceived to be of equal or greater value, (usually, but not always, a given sum of money). Facilitating functions are not a direct part of either the exchange of title or the physical movement of produce.

Standardisation: Standardisation is concerned with the establishment and maintenance of uniform measurements of produce quality and/or quantity. This function simplifies buying and selling as well as reducing marketing costs by enabling buyers to specify precisely what they want and suppliers to communicate what they are able and willing to supply with respect to both quantity and quality of product. In the absence of standard weights and measures trade either becomes more expensive to conduct or impossible altogether. In Nepal such was the diversity of weights and measures used with respect to grain within the country, that it was easier for some districts to conduct trade with neighbouring states in India than it was to do business with other districts within Nepal. Among the most notable advantages of uniform standards, are:

⦿ Price quotations are more meaningful

⦿ The sale of commodities by sample or description becomes possible

⦿ Small lots of commodities, produced by a large number of small producers, can be assembled into economic loads if these supplies are similar in grade or quality

⦿ Faced with a range of graded produce the buyer is able to choose the quality of product he/she is able and willing to purchase.

7.15 ADVANTAGES AND DISADVANTAGES OF DIVERSIFICATION IN AGRICULTURE ENVIRONMENTAL SCIENCE

Agriculture is the pre-dominant economic line of work of the rural communities in India, and plays a vital role in the socio-economic development of these communities. India acquires its major share of production revenues from the rural / agricultural sector of the economy. The agriculture sector in India is enormously significant in spite of its declining share in GDP. Sectoral shifts occurred as a result of the industrialization which had raised the Services sector shares in GDP during the 1990's, where as Agriculture, which had a major share in GDP in the 1950's, contributed only 22.5 % by the end of March 2004. Thus came up a pressing need for a paradigm shift in the government's agricultural policy to address the problems faced in the agricultural sector in the new domestic and global economic environment and avenues to enhance the income of the farmers. The possible solution for meliorating the agro sector is Diversification. This study traces the definition of diversification, area expansion problems, immediate needs, and its future prospects.

Diversification can also involve "a shift of resources from one crop (or livestock) to a larger mix of crops and livestock, keeping in view the varying nature of risks and expected returns from each crop/livestock activity, and adjusting in such a way that it leads to optimum portfolio of income". It is a way of a gradual movement from subsistence staple food crops towards diversified market-oriented crops which have a larger potential for land returns. DOA being a strategy would open up opportunities, to a large extent, for value addition in agriculture and will also lead to better crop planning and improve the earning opportunities in the farm community. In India, Andhra Pradesh has been proactive in taking up agricultural diversification as a strategy to accelerate the growth of agriculture.

7.16 CROP DIVERSIFICATION

Crop diversification takes into account the economic returns from different value-added crops. It also implies the effective use of environmental as well as human resources to grow a mix of crops with complementary marketing opportunities, and it entails shifting of resources from low value crops to high value crops. Due to globalization, crop diversification in agriculture is also a means to increase the total crop productivity in terms of quality, monetary and quantity value under specific, diverse agro-climatic situations all over the world.

7.16.1 Approaches to Crop Diversification in Agriculture

Horizontal diversification - the primary approach to crop diversification used in production agriculture. In this approach, diversification normally takes place through crop intensification which means adding new high-value crops to existing cropping systems as a way of improving the overall productivity of a particular farm or a region's farming economy as a whole.

Vertical diversification approach in which value is added to the products by farmers through various methods such as processing, regional branding, packaging, merchandising, or other efforts to enhance the product.

Opportunities for crop diversification normally vary depending upon the risk, opportunity and the feasibility of proposed changes within a socio-economic and agro-economic context. Crop diversification may occur as a result of government policies. The "Technology Mission on Oilseeds", "Spices Development Board", "and Coconut Development Board" etc is some examples where the Indian government created policies to thrust changes upon farmers and the food supply chain at large as a way of promoting crop diversity.

Crop diversification is the outcome of several interactive effects of many factors:

Environmental factors which includes irrigation, rainfall, and temperature and soil fertility.

Price-related factors which includes output and input prices with respect to national and international trade policies and other economic policies that affect the prices either directly or indirectly.

Technology-related factors which includes seeds, fertilizers and water technologies, but also those related to marketing, harvest, storage, agro-processing, distribution, logistics, etc.

Household-related factors which includes regional food traditions, fodder and fuel as well as the labor and investment capacity of farm people and their communities.

Institutional and Infrastructure-related factors which includes farm size, location and tenancy arrangements, research, in-field technical support, marketing systems and government regulating policies, etc.

All these five factors are interrelated.

7.16.2 Key Drivers of Diversification

The key drivers of diversification that are identified are : (1) Food Security; (2) Employment generation through creation of off-farm and non-farm investment opportunities within the capabilities of the resource-poor farmers; (3) Changes in crop patterns and farming systems; (4) More effective use of land and water resources; (5) Market access initiatives replacing risk aversion with risk acceptance; (6) Changing consumer demands irrespective of the nature of habitation and standards of living due to spread-effect of health consciousness caused by the visual media and non-discriminatory demand for quality goods, and (7) The role of urbanization in fast developing countries like India.

Crop diversification can better tolerate the ups and downs in the market value of farm products and may ensure economic stability for farming families of the country. The adverse effects of aberrant weather, such as erratic and scanty rainfall and drought are very common in a vast area in agricultural production of the country. Incidence of flood in one part of the country and drought in the other part is a very frequent phenomenon in India. Under these aberrant weather situations, dependence on one or two major cereals (rice, wheat, etc.) is always risky. Hence, crop diversification through substitution of one crop or mixed cropping/inter-cropping may be a useful tool to mitigate problems associated with aberrant weather to some extent, especially in the arid and semi-arid drought-prone/dry land areas.

8

DAIRY FARM MANAGEMENT

8.1 OBJECTIVES

After studying this chapter you will be able to understand:

- Digital Dairy Management
- Scope for Dairy Farming
- Importance of Dairy Farming

8.2 INTRODUCTION

Dairy farming is a class of agriculture for long-term production of milk, which is processed (either on the farm or at a dairy plant, either of which may be called a dairy) for eventual sale of a dairy product. Dairy farming has a history that goes back to the early Neolithic era, around the seventh millennium BC, in many regions of Europe and Africa. Before the 20th century, milking was done by hand on small farms. Beginning in the early 20th century, milking was done in large scale dairy farms with innovations including rotary parlors, the milking pipeline, and automatic milking systems that were commercially developed in the early 1990s.

Milk preservation methods have improved starting with the arrival of refrigeration technology in the late 19th century, which included direct expansion refrigeration and the plate heat exchanger. These cooling methods allowed dairy farms to preserve milk by reducing spoiling due to bacterial growth and humidity.

Worldwide, leading dairy industries in many countries including India, the United States, China, and New Zealand serve as important producers, exporters, and importers of milk. Since the late 20th century, there has generally been an increase in total milk production worldwide, with around 827,884,000 tonnes of milk being produced in 2017 according to the FAO.

There has been substantial concern over the amount of waste output created by dairy industries, seen through manure disposal and air pollution caused by methane gas. The industry's role in agricultural greenhouse gas emissions has also been noted to implicate environmental consequences. Various measures have been put in place in order to control the amount of phosphorus excreted by dairy livestock. The usage of rBST has also been controversial. Dairy farming in general has been criticized by animal welfare activists due to the health issues imposed upon dairy cows through intensive animal farming.

Dairying is an important source of subsidiary income to small/marginal farmers and agricultural labourers.

The manure from animals provides a good source of organic matter for improving soil fertility and crop yields.

The gober gas from the dung is used as fuel for domestic purposes as also for running engines for drawing water from well. The surplus fodder and agricultural by-products are gainfully utilised for feeding the animals.

The small/marginal farmers and land less agricultural labourers play a very important role in milk production of the country. Dairy farming is now taken up as a main occupation around big urban centres where the demand for milk is high.

8.3 SCOPE FOR DAIRY FARMING AND ITS NATIONAL IMPORTANCE

The total milk production in the country for the year 2001-02 was estimated at 84.6 million metric tonnes. At this production, the per capita availability was to be 226 grams per day against the minimum requirement of 250 grams per day as recommended by ICMR. Thus, there is a tremendous scope/potential for increasing the milk production. The population of breeding cows and buffaloes in milk over 3 years of age was 62.6 million and 42.4 million, respectively (1992 census).

8.3.1 Management of Dairy

The scheme for dairy, farming should include information on land, livestock markets, availability of water, feeds, fodders, veterinary aid, breeding facilities, marketing aspects, training facilities, experience of the farmer and the type of assistance available from State Government, dairy society/union/federation.

Technical Feasibility

1. Nearness of the selected area to veterinary, breeding and milk collection centre and the financing bank's branch.

2. Availability of good quality animals in nearby livestock market.

3. Availability of training facilities.

4. Availability of good grazing ground/lands.

5. Green/dry fodder, concentrate feed, medicines etc.

6. Availability of veterinary aid/breeding centres and milk marketing facilities near the scheme area.

Economic Viability

1. Cost of for feeds and fodders, veterinary aid, breeding of animals, insurance, labour and other overheads.

2. Output costs i.e. sale price of milk, manure, gunny hags, male/female calves, other miscellaneous items etc.

8.3.2 Farmers

Modern and well established scientific principles, practices and skills should be used to obtain maximum economic benefits from dairy farming.

Some of the major norms and recommended practices are as follows:

Housing

1. Construct shed on dry, properly raised ground.

2. Selling of the old animals after 6-7 lactations.

Feeding of Milch Animals

1. Feeding the animals with best feeds and fodders.

2. Giving adequate green fodder in the ration.

Milking of Animals

1. Milking the animals two to three times a day.

Protection Against Diseases

1. Be on the alert for signs of illness such as reduced feed intake, fever, abnormal discharge or unusual behaviour.

Breeding Care

1. Animal should be closely observed and keep specific record of its coming in heat, duration of heat, insemination, conception and calving.

Care During Pregnancy

1. Give special attention to pregnant cows two months before calving by providing adequate space, feed, water etc.

Marketing of Milk

1. Marketing milk immediately after it is drawn, keeping the time between production and marketing of the milk to the minimum.

2. Production of milk produces for better storage to give more returns

8.4 DIGITAL DAIRY MANAGEMENT

The Indian dairy industry is fraught with many difficulties such as inefficiency, deterioration of perishable food items, unsatisfactory quality of commodities, malpractices in weights and measures, mismatch of demand and supply, long waiting times, exorbitant corruption, rude behavior of shopkeepers and poor service delivery. Streamlining of SCM processes will result in increased operational efficiency, thereby reducing transit losses and pilferages.

From Cow udder to customer mouth, providing with the perfect solutions to in all operations challenges in terms of design, development procurement and production site operations, thorough understanding of production processes, building utilities and site management, plant performance will be increased by enabling team Members at every level to work with greater efficiency and high confidence

Digitalization will cover entire enterprises operations will make in to one hub, with enable IT solutions one of the most complex areas to manage is the integration of IT and staff commitment to new ways of doing business and support for smooth integration.

- Technical architecture
- Project management
- Strategic partner management
- Quality assurance
- Technology prototyping
- Business requirements management
- Enterprise architecture
- User experience design

8.4.1 Indian Dairy Business Challenges and System Streamlining

Small Holder Challenges

- Inadequate feeding of animals
- More disease incidence
- Low genetic potential of animals
- Lack of chilling capacities
- Exploitation of farmers

- High production costs
- Delayed payment of dues

Procurement Challenges

- Milk base mainly consisting of small holders
- Involvement of too many intermediaries:
- Gaps in information
- Absence of a screening system
- Lack of Infrastructure
- Manipulation of the quality of milk by the farmers

Co-Operative Challenges

- Less number of member farmers
- Lower participation in the decision making process
- Losses
- Low prices of milk
- Inefficient services
- Insufficient Infrastructure

Process Challenges

- Seasonality of production and fluctuating supply
- Absence quality standards
- Adulteration and Food safety
- Lack of trained and skilled workers

Storage and Logistics Challenges

- Lack of cold storage facilities
- Gap in the cold chain and transport facilities

Marketing Challenges

- Majority of the Market is still unorganized
- Acceptability of the Consumer base
- Less penetration to the rural Market
- Lack of transparent milk pricing System

Measuring an Organization's Total Impact Helps to Show the Best Route Forward

- Improve capital and labour productivity

- Define architecture for obtaining necessary environmental clearance for new factories and reforms on land acquisition

- Resolve IT adoption issues

- Avoid duplication of efforts with the existence of multiple stakeholders

- Reducing gaps in demand and supply

Preventing Postponement in Project Implementation

Streamlining of business processes and adoption of advanced technologies in the sector is expected to enable them to overcome such strategic hurdles to a large extent.

- Innovation

- Business process re-engineering

- Enterprise resource planning (ERP)

- Business intelligence (BI)

- Document management system (DMS)Dashboards

- RFID tagging

- Multi buyer and multi system (MBMS) system

- High-value consumer management system (HVCMS)

- Estimation and tendering systems

The common major problem of transparency and data inconsistency, Processes are generally driven by skilled people and, predictability across projects is missing. While improving processes is a natural change and a function of industry maturity, technology is being seen as a lever for making this change in a faster and more predictable manner.

8.4.2 Preserving Maximum Margins

Traditionally, manufacturing companies have pursued a blend of four approaches to preserving margin.

1. Product price increases are one way to ensure desired margins, particularly in the face of input cost inflation, but as we wrote in a Perspective last year, deciding to take a price increase and getting it to stick in the marketplace are very different. Increasing both the quality and the frequency of product innovation can be an effective way to refresh the product portfolio to "beat the fade" and recover premium margin opportunities and new product development.

2. Adjusting product mix is an approach companies have successfully employed in the past promoting higher-margin items at the expense of lower-margin

alternatives yet in today's commercial environment, where consumers are trading down to stretch disposable income and maximize perceived value, the luxury end of many manufacturers' product portfolios is not performing well.

3. Driving out unnecessary waste with cost-efficiency programs has proven to be a very effective way to preserve margin levels and has enabled companies to be leaner and more agile in responding to market forces, yet very few companies "cost cut" their way to prosperity, and there is the very real fear of marginalizing product quality of losing sight of the "total cost" picture when chasing cost reduction targets.

The most immediately available tactic for margin preservation is cost reduction, and it has been a primary focus for the vast majority of supply chain organizations. Yet effectively reducing costs can be tricky, with companies discovering that cost savings in one part of the supply chain can be more than offset by unintended consequences elsewhere.

Total dairy supply chain costs = S (Production cost + Transportation cost + Input inventory cost + Output inventory cost)

Strategic supply chain plan and operations don't go as expected results. These costs can arise as a result of disconnects between:

1. Strategic decisions, such as supply network design, inventory policy, and service-level obligations

2. Tactical decisions, such as supplier selection, factory-run strategies, or logistics partners

3. Operational decisions, such as out-of-footprint sourcing, expedited shipments, or accessorial charges

Connecting the decision-making process across strategy, tactics, and operations is important, but having visibility into these decisions and the consequences is just as important. There will always be unanticipated costs things go wrong in supply chain operations after all.

A. Identify the unplanned cost elements; collect the necessary data.

B. Analyze the data, understand the business drivers of the unplanned costs, and identify a pattern to properly assign costs.

C. Assign the costs to suppliers.

⊙ The purpose of this effort is not to penalize suppliers in the short term but to evaluate the optimal set of suppliers based on full disclosure of total cost. The end result may well be the replacement of a supplier with a more efficient alternative; however, it also may be that the driver of unanticipated costs falls at the feet of the manufacturers that can make changes to their role in the relationship that will eliminate the issues. Further, this approach is empowering to those in the supply chain organization who have a role in both identifying and reducing supply chain costs.

8.5 VISION OF DIGITAL DAIRY MANAGEMENT: FOCUS ON THREE KEY AREAS

8.5.1 Vision and Focus Areas

- ⦿ Area1 :Internet Infrastructure as a Utility for Every Dairy dept

- ⦿ Area 2:Management & Services on demand

- ⦿ Area 3:Digital Empowerment of farmer/Staff/customers

8.5.2 Internet Infrastructure as a Utility of Every Dept

- ⦿ E-Governance of (3P's) Services :(Procurement/production/promotion

Dairy Business Process Re-engineering using IT to improve transactions Form Simplification, reduction/ Online office and tracking, Interface between departments Use of online repositories, Integration of services and platforms: Receipts, Payment Gateway, Mobile Platform.

E.g. Payment, purchases, other business transactions etc.

- ⦿ Electronic Databases – all databases and information to be electronic, not manual

- ⦿ Dairy Grievance Redresses -By using IT to automate, respond, analyse data to identify and resolve persistent problems – largely process improvements

The key business drivers for investment in technology surveillance are

Regulatory Drivers

- o   Near-time and real time monitoring capabilities

- o   Horizontal Trade Surveillance Model

- o   Emphasis on pre-emptive controls and robust trade surveillance Platforms

- o   Global Legal Entity Identifier (LEID) data hierarchy requirements

Technology Drivers

- o   Measurement and monitoring of unstructured data

- o   Collective screening of unstructured data and structured data for complete trade reconstruction

- o   Reduction in false positives

- o   Improved visualization for senior management decisions

Operational Drivers

o   Record-keeping requirements: maintaining historical data over a period of time

o   Blind spots between disparate systems

o   Sourcing, capturing, and maintaining a data ecosystem for analysis

o   Surveillance across Chinese walls

The overall project focuses on advanced methodologies to guarantee the Dairy business, and new technologies for improvement of competitiveness, sensory quality, and for the maximization of customer satisfaction and environmental protection.

The competitive advantages of companies can stimulate and support collaboration. Competitive advantage is the ability of a company to protect itself against its competitors. Mainly five factors contribute to the competitive ability: competitive pricing, premium pricing, value-to-customer quality, reliability of delivery and innovation in production

1. Procurement AMCU operations
2. Plant process efficiency
3. Innovative systems for packaging
4. Measurement and control of product quality
5. E-Marketing
6. ICT-based logistics platform
7. Data accuracy
8. Validation of the prototype

1. E-Governance –of Procurement with advanced technology Daily milk collection system through AMCU: (From agent level collection)

The initiative of installing the AMCUS – Automatic Milk Collection Unit Systems at village level to enhance the transparency of transaction between the farmer and the Dairy, These system not only ensured the transparency but also gave a unique advantage by reducing the processing time to 10 percent of what it used to be prior to this. Indeed got the entire supplier information through the systems integration. The Information related to members, fat content, volume of the milk procured and the amount payable to the member are accessible to the members in the form of a database.

The benefits of the AMCUS system:

◉ The rural people are getting benefited much by the IT initiatives,

◉ The benefits of various projects such as DISK are yet to be realized.

◉ The following are the demonstrated benefits of the ICT platform.

- ⦿ Time reduction
- ⦿ Reduction of pilferage
- ⦿ Reduced human errors
- ⦿ Transparent quality and qty analysis
- ⦿ Wastage is reduced
- ⦿ Transparency of operations and payments
- ⦿ Total Operational integration

Advantages of AMCUS

1. Saving in quantity of sample milk
2. Saving of chemicals and detergents
3. Saving of expenditure on glassware
4. Saving in stationery and time
5. Saving in expenditure on staff.
6. Gaining confidence of milk producers

Through transparent system and increased, Milk procurement quality and hygiene

8.5.2 Automation of Manufacturing Plants

- ⦿ Process flow automation :
- ⦿ The need for technology up gradation in the infrastructure sector, to reduce gestation lags and improve the quality of products that can maintain the balance between sustainability and development is more than ever.

Automatic Process Monitoring

The use of digital imaging technology for the automatic monitoring of dry sugar granules and powders, this system provides particle size data to production line operators for process control and product quality improvement.

Measurement and Control of Product Quality

Prototypes of devices (sensors) will be developed to measure key parameters for monitoring and maintaining the quality of transported and stored products, but also useful in tracing the productive chain. Innovative biological physical technologies will be applied to protect the products after harvesting and during transport/storage.

8.5.3 Measurement and Control of Product Quality

Prototypes of devices (sensors) will be developed to measure key parameters for monitoring and maintaining the quality of transported and stored products, but also useful in tracing the productive chain. Innovative biological physical technologies will be applied to protect the products after harvesting and during transport/storage.

Regular Utilities

- Monitor energy usage
- Reduce energy costs
- Diversify energy sources
- Monitor water quality
- Reduce effluents
- Produce energy from
- Waste biomass
- Produce green energy

CIP Station

- Reduce cleaning downtime
- Recover water – mix products
- Clean multiple equipment at the same time
- Ensure equipment cleaning and sanitizing
- Reduce water consumption
- Reduce chemical usage

Plant Management

- Reduce Inventories
- Measure and reduce product and energy loss
- Download work orders from ERP system
- Improve processes on-line
- Improve HACCP system
- Implement global production tracking
- Manage products and materials genealogy
- Manage Process and Machine safety
- Produce healthy and affordable products
- Reduce packaging costs and materials
- Maximize performance
- Flexible line changes
- Track and traceability

Innovative systems for packaging: New packaging will be designed considering the final destinations; respecting the existing standards; with technical, structural and micro structural characteristics calibrated depending on the product, ensuring the best possible hygiene, with an economic and environmental impact as limited as possible, and packaging to extend the shelf- life and improve the "perceived quality" and natural aromatic component of food products;

The project involves the introduction of the concept of excellence of with innovative production processes and improved packaging that stabilize or maintain the quality of dairy food products throughout their conservation.

8.5.4 Innovation in Dairy Packing

- Convenience
- Healthier foods
- Reduced portions
- Natural ingredients
- Energy boosting drinks

Various Packing Categories

- Different size packages /Multi-packs
- Variety packs /Retail ready displays
- Sustainable packaging

Consumer Choice Based Packing

- Low fat / Fat-free / Reduced sodium and sugar / Lactose-free /Organic /
- Probiotic cultures, fortified with nutrients, calcium and protein

Process efficiency: Infrastructure processes are complex and require strong technical and functional skills. Being an industry which is highly labour intensive, throughput of work that gets completed on a daily, weekly and monthly basis will be strongly aided by technology. Unlike conventional manufacturing industry ERPs, this industry requires several real-time, mobility and analytical solutions which will get more out of the individual.

Data accuracy: Trusting data that originates from a site, a toll plaza, earthmoving equipment in use is always a challenge. Manual representation in any format always has room for data getting adulterated, unconsciously or otherwise. This has eroded trust in the infrastructure space. The social perception for this industry is therefore taking a beating. Eliminating manual intervention, capturing the moment of truth and using more decision support systems will certainly create a strong connect with both the investor as well as the user community.

Scalability: While scalability is usually associated with volume, in the infrastructure industry it needs to get associated with heterogeneity. Hence, technology needs to transcend the boundary of infrastructure and look at how financial services, the concept of enterprise architecture, use of social analytics, machine-to-machine interaction as well as the internet of things can certainly be explored.

Achieving Energy Efficiency

Sustainable solutions are divided into four stages:

1. Measure to identify sources of energy savings or malfunctions

2. Install energy saving systems and equipment

3. Improve long-term use through control system management, support and training tools while maintaining a high level of performance

4. Analyze gains through ongoing maintenance, supervision and controls

E-Marketing Technology SCM System of Delivery of Goods

⊙ All agents connected with broadband

⊙ Free Wi-Fi to all agents / retailers

⊙ Digital Literacy program for Dealers/ agents/marketing retailers

Technology for Healthy E-Marketing

⊙ Online sales consultation

⊙ Online milk and product sales

⊙ Online service Cold chain parlours should establish in market coverage area

⊙ Milk e- wallets, e- payments

Technology for Planning

⊙ Technology based decision making

⊙ Dairy GIS Mission Mode Project

ICT-Based Logistics Platform

⊙ It will standardize as far as possible, the use of hardware and software technologies to reduce delivery times; to allow continuous monitoring and real-time recording of parameters that are critical for storage of products; to ensure compliance with mandatory rules of hygiene and traceability of the productive chain, fight against waste (failed deliveries) and increase the number of satisfied customers.

⊙ Strategic supply chain plan and operations don't go as expected results. These costs can arise as a result of disconnects between:

1. Strategic decisions, such as supply network design, inventory policy, and service-level obligations

2. Tactical decisions, such as supplier selection, factory-run strategies, or logistics partners

3. Operational decisions, such as out-of-footprint sourcing, expedited shipments, or accessorial charges

4. Connecting the decision-making process across strategy, tactics, and operations is important, but having visibility into these decisions and the consequences is just as important. There will always be unanticipated costs things go wrong in supply chain operations after all.

⊙ Identify the unplanned cost elements; collect the necessary data.

⊙ Analyze the data, understand the business drivers of the unplanned costs, and identify a pattern to properly assign costs.

8.5.5 Assign the Costs to Suppliers

The purpose of this effort is not to penalize suppliers in the short term but to evaluate the optimal set of suppliers based on full disclosure of total cost. The end result may well be the replacement of a supplier with a more efficient alternative; however, it also may be that the driver of unanticipated costs falls at the feet of the manufacturers that can make changes to their role in the relationship that will eliminate the issues. Further, this approach is empowering to those in the supply chain organization who have a role in both identifying and reducing supply chain costs.

Validation of the Prototype

The program will conclude with the development and validation of a prototype logistics platform, which will be based on all the results obtained by simulation approaches. A qualitative and quantitative assessment of the benefits in terms of reliability, reproducibility, safety, effectiveness and efficiency (cost reductions) will also be carried out. The program will end with the completion of the prototype of logistics platform.

The overall project focuses on advanced methodologies to guarantee the Dairy business and new technologies for the improvement of competitiveness, sensory quality, and for the maximization of customer Satisfaction and environmental protection.

Admin/Management for All

⊙ Online Hosting of Information and documents

    o   Open, easy access to information

    o   Open data platform

    o   All payments and receipts through cashless transactions

- Dairy pro-actively engages through social media and web based platforms to inform customers
- 2-way communication between farmer/customer and Dairy
- Online messaging to customers on special occasions/programs
- Largely utilization of existing infrastructure – limited additional resources needed.

As a final point: The availability of safe, sustainable and healthy food is increasingly becoming a major issue, given the continued growth of the world population and the increasing environmental and sustainability concerns. The need to meet consumer demands in terms of high quality products, healthy diets, and affordable prices, fair and equitable relations in the Dairy supply chain, food safety, and ethics of productions are a huge challenge. In addition, climate change, competition for agricultural land use and shifting dietary habits are constantly changing the supply/demand scenarios. In order to make the food supply chain systems safer, more performing, resilient, efficient and fair, it is necessary to take account of the links between product quality, health and welfare of customers and issues related to the exploitation of ecosystems, sustainability and biodiversity.

Protecting the dairy value chain is a target that can be achieved only by a multi disciplinary effort, taking into account the different stages of the dairy production chain.

Ex: The costs can arise as a result of disconnects between:

1. Strategic decisions, such as supply network design, inventory policy, and service-level obligations
2. Tactical decisions, such as supplier selection, factory-run strategies, or logistics partners
3. Operational decisions, such as out-of-footprint sourcing, expedited shipments, or accessorial charges

The application of new enabling technologies within the Supply Chain Integrated Approach could help redefine the dairy- food scenario will reshape the value chain towards advanced scenarios in which all the enterprises benefit from better management processes and new products and services. In fact, a strict control of raw materials could help detect all risks and act quickly in critical phases (using appropriate corrective measures to prevent the processing of non-conforming products), with obvious advantages for all the operators involved in the entire supply chain.

In addition, these new integrated technologies, allow the creation of an "intelligent" scenario of systems and devices capable of interacting with all operators to significantly improve the level of knowledge and management processes related to product quality and safety. As regards foods safety more specifically, a comprehensive approach (from primary production to the final consumer) to Supply Chain Integration could ensure its effective optimization and improvement throughout the supply chain in which therefore distinguishes it on national and international markets.

9

CROP MANAGEMENT

9.1 OBJECTIVES

After studying this chapter you will be able to understand:

- Preparation of Soil

- Post Harvest Storage

- Impact of Climate Change on Agriculture

- Post Insects and Climate Change.

9.2 INTRODUCTION

The group of agricultural practices used to improve the growth, development, and yield of agricultural crops. The combination, timing, and sequence of the practices used depend on the biological characteristics of the crops (whether winter or spring crops), the harvested form (grains, green feed, and so on), the sowing methods (row, nest, or wide-row), the age of the plants, and the soil, climatic, and weather conditions.

The principal crop-managing practices vary according to the class of crops. Winter crops require autumn topdressing with mineral fertilizers to improve winterhardiness of the plants, snow retention, spring topdressing, and harrowing. Solid-planted spring crops call for harrowing and topdressing; in arid regions they additionally need soil packing after planting. For row crops in preemergence, the crust of the soil must be broken up with harrows or rotary hoes; interrow tillage, blocking, thinning of sprouts, and topdressing are also indicated in the postemergence period. Perennial grasses require harrowing in the spring, harrowing after mowing, and topdressing.

Special crop-managing practices for individual crops include hilling, suckering, pinching, and chopping. Other field practices include crop irrigation and mechanical, biological, and chemical methods of combating weeds, pests, and diseases.

Agriculture is the science, which mainly deals with the diverse processes or the methods used for the cultivating different varieties of plants and livestock farming or animal husbandry on the basis of human requirements.

When plants of the same variety are cultivated on a large scale, they are called crops. The crops are divided on the basis of the seasons in which they grow:

Kharif Crops-These crops are sown in the early monsoon season, which generally varies by crop and region of cultivation. In India, Kharif crops are sown at the beginning of the rainy season, between the month of June and July. These crops are harvested at the end of monsoon season, between the month of September and October. Paddy is the main Kharif crop.

Rabi Crops- These crops are sown during winter and after the monsoon, which is between the month of October and November. In India, Rabi crops are harvested during the spring between the month of March and April. Wheat is the main Rabi crop.

About 70% of the Indian population practices agriculture. Hence, the production and management of crops is an important aspect to ensure optimal productivity in the fields. The major agricultural practices involved in crop production and management are listed below:

- ◉ Preparation of Soil
- ◉ Sowing of Seeds
- ◉ Addition of Manure and Fertilizers
- ◉ Irrigation
- ◉ Protection from Weeds
- ◉ Harvesting
- ◉ Storage

9.3 PREPARATION OF SOIL

The soil is loosened and tilted before the seeds are sown. Ploughs are used for the purpose. If the soil contains big lumps, they are broken with the help of a hoe. This process aerates the soil so that the roots breathe easily. The nutrients and minerals get properly mixed with the soil and come at the top. Thus, the fertility of the soil increases and is fit for plantation.

9.3.1 Sowing of Seeds

The good quality, infection-free seeds are collected and sown on the prepared land. The seeds should be sown at proper depths and proper distances. Following are the various methods used to sow the seeds:

- ⊙ Traditional techniques
- ⊙ Broadcasting
- ⊙ Dibbling
- ⊙ Drilling
- ⊙ Seed dropping behind the plough
- ⊙ Transplanting
- ⊙ Hill dropping
- ⊙ Check row planting

9.3.2 Addition of Manures and Fertilizers

The soil may not have the right nutrients to efficiently sustain plant growth. Hence, manures and fertilizers are added to the soil to increase its fertility and help plants grow better. Manure is prepared by using decomposing plant and animal matter in compost pits. Fertilizers, on the other hand, are chemicals prepared in factories which contain nutrients for a specific plant. They give faster results than manures. However, when excessively used, they turn the soil infertile.

9.3.3 Irrigation

Crops require water at regular intervals for proper growth. The supply of water to the plants is known as irrigation. Well, rivers, lakes, tube-wells are different sources for irrigation. The traditional methods of agriculture involve the use of humans and animals. The various traditional ways are moats, chain-pump, dhekli, rahat.

The modern techniques of irrigation include the sprinkler system and the drip system. Water is very important for the germination of seeds. It helps in the proper development of flowers, fruits, seeds, and plants. Therefore, it should be present in plants in large quantities.

9.3.4 Protection from Weeds

The undesirable plants that grow along with the crops are called weeds. These weeds, feed on the nutrients provided to the crops and thus reduce the supply of nutrients to the crops, thereby, inhibiting their growth. The growth of these weeds needs to be prevented in order to enhance the growth of the plants.

The process of removal of weeds is called weeding. To achieve this, weedicides are employed, which are essentially chemicals specifically made to destroy weeds. They are usually sprayed before seeding and flowering.

9.3.5 Harvesting

When the crop matures, it is cut for further processing. This process is known as harvesting. It is usually manual labour, done with the help of sickle. However, mechanical harvesting is used these days – machines such as combine harvesters are used where the crops are harvested and threshed in one go.

- ⊙ Threshing- Separation of grains from the harvested crops is called threshing. It is done either mechanically or by cattle.

- ⊙ Winnowing- The separation of grains and chaff is called winnowing. It is done either mechanically or manually.

9.3.6 Storage

The grains should be properly stored if they are to be kept for longer periods. They need to be protected from pests and moisture. The freshly harvested seeds should be dried before they are stored. This prevents the attack from microorganisms and pests.

The harvested and separated grains are stored in airtight metallic bins or in the jute bags. Dried neem leaves are added to protect them from damage at home. Large amounts of grains are stored in granaries or silos with specific chemical treatments, to protect them from pests and insects.

9.3.7 Food from Animals

Animals are an important source of food. The rearing of animals for food is known as animal husbandry. Some animals like cows and buffaloes are reared for milk, others for meat like goats and poultry. Some people consume fish as a part of their diet. Honey bees are reared for honey. Thus, animals are an integral source of food and food products.

9.3.8 Importance of Crop Management

Adoption of best crop management practices improves crop productivity and can contribute to greater yields with improved quality. Crop management is the set of agricultural practices performed to improve the growth, development and yield of crops. It begins with a seedbed preparation, sowing of seeds and crop maintenance; and ends with crop harvest, storage and marketing. The timing and sequence of agricultural practices depend upon several factors, such as winter or spring crops; harvested products such as grain, hay and silage; sowing methods-broadcast and row-crops; and, plants age, soil, climate and weather conditions.

9.3.9 Seedbed Preparation

Seedbed preparation is the first step to improve crop growth and development. The ideal seedbed is uniformly firm, has adequate soil moisture near the surface, and is free from competing weeds. "Good seed-to-soil contact required" is a phrase commonly seen on seeding documents. Seed germination is improved if seeds have good contact with soil. However, too firm a seedbed makes it challenging to get the seed into the ground.

The two primary methods of seedbed preparation are conventional tillage and reduced or no-tillage. The traditional conventional tillage involves turning over the entire plow depth and exposing large quantities of soil organic matter to oxidation. However, reduced or no-tillage practices can lead to an accumulation of soil carbon, which can ultimately benefit soil health and improve crop yields in the long run.

9.3.10 Planting

After the seedbed has been prepared, seed should be sowed 1.5 to 2.0 inches deep to ensure proper moisture availability for good seed germination. The seed requires optimum moisture and temperature conditions to germinate, so always pay close attention to soil temperature and moisture requirements for proper seed germination.

9.3.11 Fertilization

Fertilization may be an important component of crop management. Soils should be tested for available plant nutrients before adding fertilizers to any crop. The addition of appropriate fertilizers determined from the soil and/or plant analysis can ensure the planted crop's nutritional requirement.

The amount of fertilizer, type (bulk-blended or mixed), forms (gas, dry solids or liquid), timing, and method of application (broadcast, deep placement, dribble, foliar, starter, post-emergent, row, strip and variable rate), are all determined by a variety of factors, such as crop and fertilizer type, soil and weather conditions. The previous crop (legume) and past manure applications also influence crop nutrient needs. So, past manure applications should always be accounted for in determining crop needs.

9.3.12 Pest Management

Pest management is another important aspect of crop management. Pesticides can be powerful tools for controlling pests in most crops, mainly if used correctly based upon specific pest species. Additionally, integrated pest management (IPM) practices can provide growers with an economical option that is safer and often more beneficial to human and natural resources. This IPM approach incorporates mechanical, biological and chemical (labeled pesticides) pest control methods.

Using the same active ingredient repeatedly on the same piece of land, regardless of the product name, will cause pests to develop resistance over time. This makes the chemical less useful or even useless over time. Thus, to avoid the development of resistance among pests, limit using the same pesticides and choose products from different chemical classes, or vary modes of action. It is best to include some cultural practices (crop rotation, companion crops) and biological controls (predators, parasitoids) to avoid the development of pest resistance to pesticides. Generally, diverse cropping systems tend to decrease the probability of widespread crop failures and pest pressure, while improving soil quality and crop yields. The crop should also be monitored regularly for any specific needs, such as nutrient deficiencies, pest outbreaks, etc., throughout the growing season.

9.3.13 Irrigation

Irrigation is another critical factor for crop production that influences the final crop yields and quality, especially in our dryland region. Over-irrigation results in leaching of nutrients to the groundwater and/or wasting water and soil erosion via surface runoff. These losses will reduce the efficiency of fertilizers, especially nitrogen.

Before you plant any crop, obtain information regarding water needs and the critical growth stages of that crop, and then determine the irrigation system efficiency to schedule irrigation. If feasible, use irrigation systems that give improved water use efficiencies, such as micro-sprinklers, low-elevation sprinklers and drip (85-95% efficient), or low- and high-pressure center pivots (75-90% efficient). In general, the flood irrigation system is less efficient (20-50%) than other methods. Additionally, if possible, schedule your irrigation during the early morning or late evening to avoid water losses via evaporation.

9.3.14 Harvesting

Finally, the yield and quality of crops depend upon the harvest management strategy. Too wet or snowy conditions can delay the harvest of the crop. High moisture content delays the mechanical harvesting (windrowing or swathing, direct combining) of the crop/seed. Most of the grain/seed crops should be harvested when they have reached the harvest maturity stage. This timing reduces the yield loss via shattering and lodging. Therefore, missing the right time to harvest often results in severe yield loss.

The maturity stage at forage harvest is a critical factor influencing the forage quality and end-use. If the forage harvest is delayed for maximum yields (for instance, alfalfa), then forage quality will deteriorate or fall below the needed optimal quality. The maximum yield of alfalfa forage is achieved at the full flowering stage; however, forage quality is highest before flowering.

9.4 POST-HARVEST STORAGE

The post-harvest storage conditions also influence the crop's forage and grain quality. The harvested crop should be stored at the proper recommended moisture content for each crop to maximize the quality, reduce pest infestation, and avoid deterioration during storage. For example, cereals stored at 14.5 percent moisture content are highly susceptible to quality loss, mold growth and insect infestation. Alfalfa forage should be baled when the moisture content is 18-20% for better quality.

9.4.1 Additional Practices

Some additional best practices to increase crop productivity and farm profitability are:

A.   Increase crop diversity

B.   Enhance beneficial pollinators population

C.   Use better weed control measures to increase harvest efficiency, crop quality and yield

D.   Improve soil quality by following the best soil management practices

E.   Add nutrients based upon availability from soil and crop needs

F.   Manage labor and input costs

G.   Keep track of all expenses and profits

H.   Keep good records to help manage a profitable farm business

I.   Engage in creative marketing

9.4.2 Climate Change Effect on Crop Productivity

Climate change and agriculture are interrelated processes, both of which take place on a global scale. Climate change affects agriculture in a number of ways, including through changes in average temperatures, rainfall, and climate extremes (e.g., heat waves); changes in pests and diseases; changes in atmospheric carbon dioxide and ground-level ozone concentrations; changes in the nutritional quality of some foods; and changes in sea level.

Climate change is already affecting agriculture, with effects unevenly distributed across the world. Future climate change will likely negatively affect crop production in low latitude countries, while effects in northern latitudes may be positive or negative. Climate change will probably increase the risk of food insecurity for some vulnerable groups, such as the poor. Animal agriculture is also responsible for CO

A greenhouse gas production and a percentage of the world's methane, and future land infertility, and the displacement of local species.

Agriculture contributes to climate change both by anthropogenic emissions of greenhouse gases and by the conversion of non-agricultural land such as forests into agricultural land. Agriculture, forestry and land-use change contributed around 20 to 25% of global annual emissions in 2010.

A range of policies can reduce the risk of negative climate change impacts on agriculture and greenhouse gas emissions from the agriculture sector.

Agriculture and fisheries are highly dependent on the climate. Increases in temperature and carbon dioxide (CO_2) can increase some crop yields in some places. But to realize these benefits, nutrient levels, soil moisture, water availability, and other conditions must also be met. Changes in the frequency and severity of droughts and floods could pose challenges for farmers and ranchers and threaten food safety. Meanwhile, warmer water temperatures are likely to cause the habitat ranges of many fish and shellfish species to shift, which could disrupt ecosystems. Overall, climate change could make it more difficult to grow crops, raise animals, and catch fish in the same ways and same places as we have done in the past. The effects of climate change also need to be considered along with other evolving factors that affect agricultural production, such as changes in farming practices and technology.

Crops grown in the United States are critical for the food supply here and around the world. U.S. farms supply nearly 25% of all grains (such as wheat, corn, and rice) on the global market. Changes in temperature, atmospheric carbon dioxide (CO_2), and the frequency and intensity of extreme weather could have significant impacts on crop yields.

For any particular crop, the effect of increased temperature will depend on the crop's optimal temperature for growth and reproduction. In some areas, warming may benefit the types of crops that are typically planted there, or allow farmers to shift to crops that are currently grown in warmer areas. Conversely, if the higher temperature exceeds a crop's optimum temperature, yields will decline.

Crop yield analysis, spatial analysis, and agricultural systems analysis are the three main approaches for studying the "Implications of a Global Climatic Warming for Agriculture," according to Smit, Ludlow, and Brklacich (1988). Crop yield analysis estimates the effects of altered environments on crop productivity levels and has been employed widely in climatic impact assessments. Spatial analysis examines the implications of climatic warming on the area and location of lands suitable for agricultural production. Agricultural systems analysis assesses the impacts of climatic change on multiple agricultural activities and on the functioning of the agrifood sector, including prices, trade pattern, and employment.

In Climate Change and World Agriculture, Parry (1990) sketches a broad picture of the effects of climate change in the chapter "Effects on Plants, Soil, Pests and Diseases." He argues that the effects of carbon dioxide (CO_2) enrichment, without associated changes in climate, would probably be beneficial for agriculture. Higher temperatures, however, could increase the rate of microbial decomposition of organic matter, adversely affecting soil fertility in the long run. Also, studies analyzing the effects on pests and diseases suggest that temperature increases may extend the geographic range of some insect pests currently limited by temperature.

Rosenzweig and Liverman (1992) compare temperate and tropical regions in "Predicted Effects of Climate Change on Agriculture." The regions differ significantly, both in the biophysical characteristics of their climate and soil and in the vulnerability of their agricultural systems and people to climate change. An analysis of the biophysical impact of climate changes associated with global warming shows that higher temperatures generally hasten plant maturity in annual species, thus shortening the growth stages of crop plants. Global estimates of agricultural impacts have been fairly rough to date, because of lack of consistent methodology and uncertainty about the physiological effects of CO_2. Climate change scenarios that do not include the physiological effects of CO_2 predict a decrease in estimated national production, but including the physiological effects of CO_2 mitigates the negative effects. Tropical regions appear to be more vulnerable to climate change than temperate regions.

Easterling et al. (1993) use another crop modeling technique, the Erosion Productivity Impact Calculator (EPIC), to determine the relationship between climate and crop growth.

The most important ecosystem service delivered by agriculture is the provision of food, feed and fibres.

The extent to which this provisioning depends on external production inputs is a fundamental issue. Agricultural ecosystems have evolved under human management. To obtain greatest possible production from the landscape, agricultural communities have developed and maintained ecosystems at their early succession state. The human selection pressure has favoured readably harvestable crops with high net production and it has penalized biomass production and accumulation on the landscape.

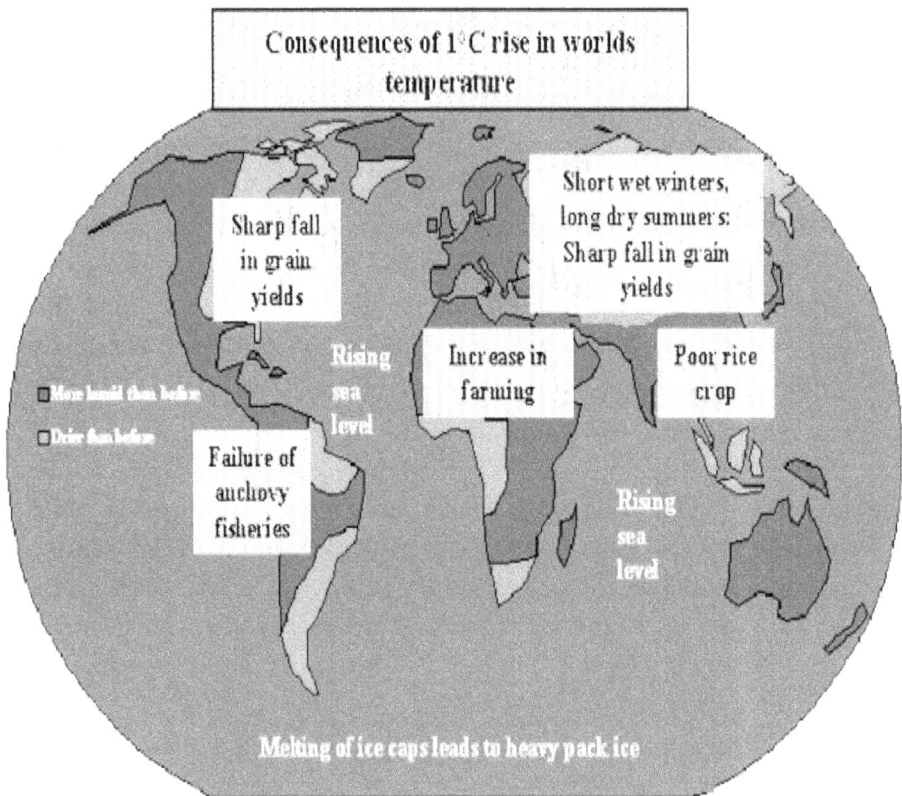

Fig. 1. Climate change effect on crop productivity.

Since the Green Revolution, mainstreamed agriculture has mainly involved controlling crop varieties and their genetics; soil fertility through the application of chemical fertilizers; and pests with chemical pesticides. The impact of this form of agriculture on the environment has been severe. There has been a significant simplification and homogenization of the world's ecosystems. Maize, wheat, rice and barley, which were once rare plants, have become the dominant crops on earth and staples in human diets.

Soil degradation is another critical concern. In agricultural ecosystems depleted of soil organic carbon, it will be increasingly difficult to produce higher yields. Each year, soil erosion destroys 10 million hectares of cropland. Forty percent of this loss is due to tillage erosion. In soils that have already experienced significant losses of soil organic matter, increased fertilization does not usually generate a net sink for carbon, because the production, transport and application of fertilizer releases higher amounts of carbon dioxide.

The FAO 'Save and Grow' model of sustainable crop production intensification calls for a 'greening' of the Green Revolution to achieve the highest possible productivity by unit of input within the ecosystem's carrying capacity. This can be achieved through the use of good quality seeds and planting materials of well-adapted varieties; a diverse range of crop species and varieties grown in associations, intercrops or rotations; the control of pests through integrated pest management; and the use of conservation agriculture and sustainable mechanization to maintain healthy soils and manage water efficiently. Greater access to technological innovations and a sound understanding of agricultural ecosystems will allow farmers to work 'smarter not harder' and work in tandem with biogeochemical processes inherent in diverse and complex ecosystems.

The FAO model for the sustainable intensification of crop production is the cornerstone of climate-smart agriculture. It guides all climate-smart strategies aimed at overcoming the inefficiencies that are responsible for yield and productivity gaps. In each crop system, there exist many climate change adaptation and mitigation options to close yield gaps and minimize the harmful environmental impacts of crop production. Options will vary among farmers and will depend on each farmer's coping and adaptive mechanisms, and the degree to which each specific climate factor is responsible for the yield and productivity gap. The solutions identified should always be cost-effective and profitable for farmers and responsive to markets. Since most technologies have both advantages and disadvantages, trade-offs will need to be made. Ensuring that these trade-offs are properly assessed demands comprehensive capacity development for all stakeholders. In particular, farmers must manage the foreseen business risks of changing their production practices (e.g. costs, investments and future value of the investments); consider the financial returns related to adapting to changes in local climate; evaluate the implications of local climate on local prices and markets; and anticipate the consequences climate change may have on crop prices in international markets.

9.5 IMPACT OF CLIMATE CHANGE ON AGRICULTURE

Global food security relies on both sufficient food production and food access, and is defined as a state when:

'all people, at all times, have physical and economic access to sufficient, safe, and nutritious food to meet their dietary needs and food preferences for an active and healthy life'.

The principal barrier to food security is currently food access. Sufficient food is produced globally to feed the current world population, yet more than 10% are undernourished.

Climate change is likely to contribute substantially to food insecurity in the future, by increasing food prices, and reducing food production. Food may become more expensive as climate change mitigation efforts increase energy prices. Water required for food production may become more scarce due to increased crop water use and drought. Competition for land may increase as certain areas become climatically unsuitable for production. In addition, extreme weather events, associated with climate change may cause sudden reductions in agricultural productivity, leading to rapid price increases. For example, heat waves in the summer of 2010 led to yield losses in key production areas including: Russia, Ukraine and Kazakhstan, and contributed to a dramatic increase in the price of staple foods. These rising prices forced growing numbers of local people into poverty, providing a sobering demonstration of how the influence of climate change can result in food insecurity.

The consensus of the Intergovernmental Panel for Climate Change (IPCC) is that substantial climate change has already occurred since the 1950s, and that it's likely the global mean surface air temperature will increase by 0.4 to 2.6°C in the second half of this century (depending on future greenhouse gas emissions). Agriculture, and the wider food production system, is already a major source of greenhouse gas emissions. Future intensification of agriculture to compensate for reduced production (partly caused by climate change) alongside an increasing demand for animal products, could further increase these emissions. It's estimated that the demand for livestock products will grow by +70% between 2005 and 2050.

While gradual increases in temperature and carbon dioxide may result in more favourable conditions that could increase the yields of some crops, in some regions, these potential yield increases are likely to be restricted by extreme events, particularly extreme heat and drought, during crop flowering. Crop production is projected to decrease in many areas during the 21st century because of climatic changes. This is illustrated in figure 1 which summarises average crop yield projections across all emission scenarios, regions, and with- or without-adaptation by farmers, showing an increasing trend towards widespread yield decreases.

Heat waves (periods of extreme high temperature) are likely to become more frequent in the future and represent a major challenge for agriculture. Heat waves can cause heat stress in both animals and plants and have a negative impact on food production. Extreme periods of high temperature are particularly harmful for crop production if they occur when the plants are flowering – if this single, critical stage is disrupted, there may be no seeds at all. In animals, heat stress can result in lower productivity and fertility, and it can also have negative effects on the immune system, making them more prone to certain diseases.

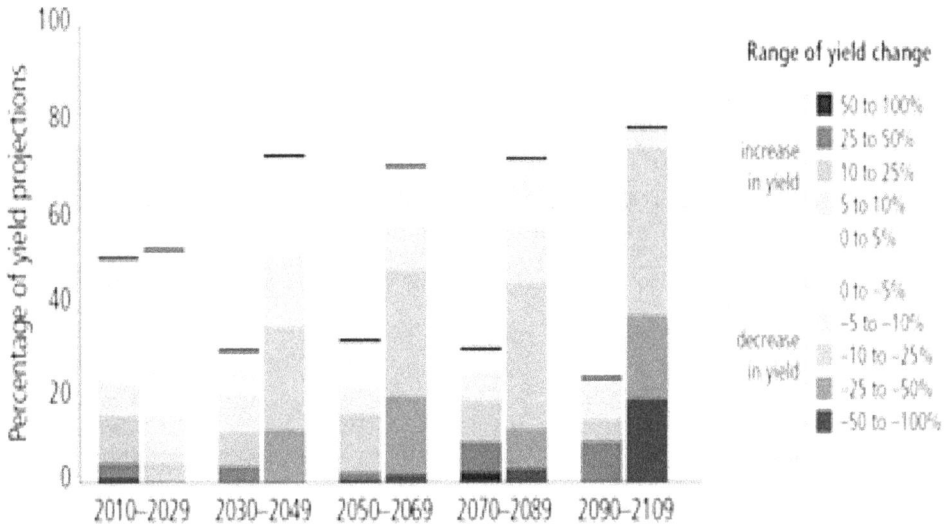

Fig. 2. Summary of projected changes in crop yields, due to climate change over the 21st century. The figure includes projections for different emission scenarios, for tropical and temperate regions, and for adaptation and no-adaptation cases combined. Relatively few studies have considered impacts on cropping systems for scenarios where global mean temperatures increase by 4°C or more. For five timeframes in the near term and long term, data (n=1090) are plotted in the 20-year period on the horizontal axis that includes the midpoint of each future projection period. Changes in crop yields are relative to late-20th-century levels. Data for each timeframe sum to 100%.

Evidence for an increase in heat waves exists from warming that has already occurred, and greater than expected increases in heat wave frequency and magnitude (figure 3). It is difficult to make accurate predictions about the future frequency and magnitude of heat waves, but there is consensus among projections that measurements for both will continue to increase in the UK, in Europe, and at a global scale. The impact of heat waves are expected to be non-uniform, with disproportionately negative effects in less developed countries. Together with other aspects of climate change such as increased drought incidence, they may exacerbate existing issues around food security.

Projected changes in climate are not limited to increases in temperature and heat waves; large changes in rainfall patterns are also expected to occur. While some regions are likely to suffer from more droughts in the future, other regions are expected to face the opposing issues of torrential rains and increased flooding. In coastal areas, rising sea levels may result in complete loss of agricultural land. Warmer climates may also lead to more problems from pests and diseases, and shifts in the geographical distribution of certain pests. For example, insects that serve as a vector for disease transmission are likely to migrate further pole-ward in the future, where livestock have so far not been exposed to these diseases.

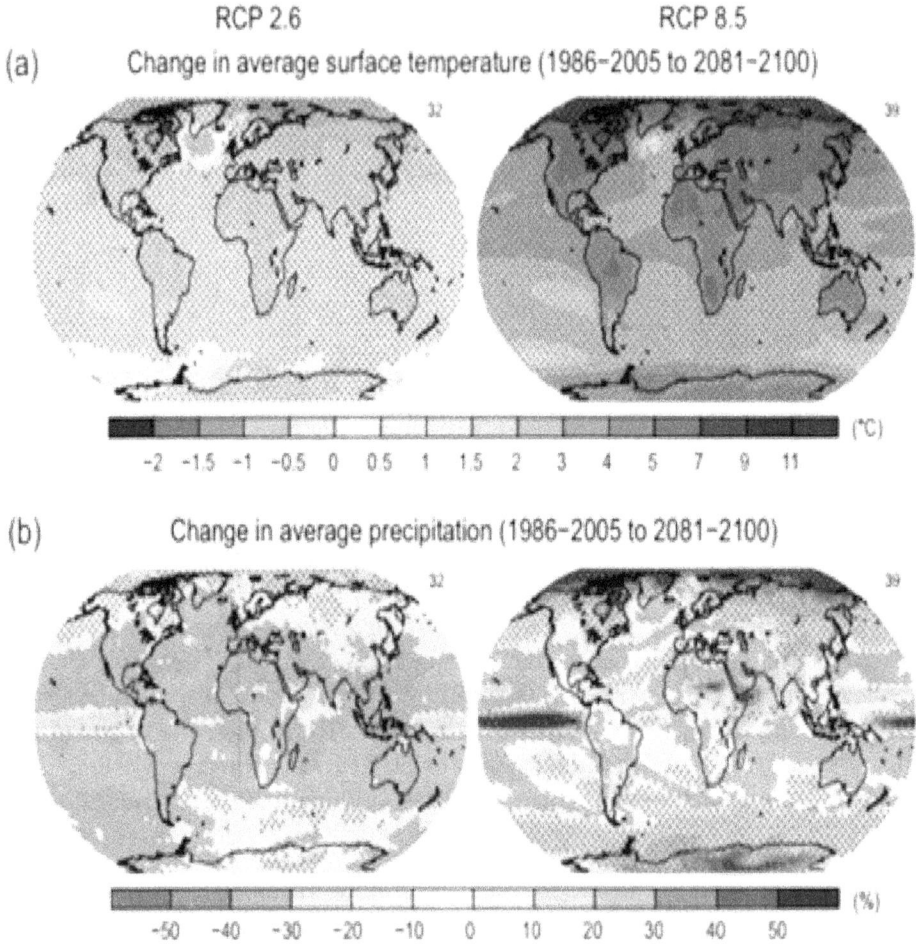

Fig. 3. Maps of CMIP5 multi-model mean results for the scenarios RCP2.6 and RCP8.5 in 2081–2100 of (a) annual mean surface temperature change, (b) average percent change in annual mean precipitation.

The responses of yield to various stresses have been well defined through experimentation in many crops. Quantifying these responses, and identifying when agriculture is most vulnerable to stress, is beneficial in helping to identify the most efficient strategies for adaptation. Crop-level adaptation to climate change is expected to be key in minimising future yield losses and may involve: changing crop cultivars, sowing time, cultivation techniques, and/or irrigation practices. Ongoing research is addressing the challenges of maintaining and/or increasing crop production under global change. Some risks to crop production from climate change and extreme weather events have been identified, and strategies suggested to help maintain production. These include: restoring farm type, crop, or cultivar scale diversity into food systems, to improve their resilience and making crop improvements that enhance stress tolerance. Other strategies may include developing pre-defined, international responses to food shortages in order to prevent food price shocks that might reduce people's access to food.

In the long run, the climatic change could affect agriculture in several ways :

- ◉ Productivity, in terms of quantity and quality of crops

- ◉ Agricultural practices, through changes of water use (irrigation) and agricultural inputs such as herbicides, insecticides and fertilizers

- ◉ Environmental effects, in particular in relation of frequency and intensity of soil drainage (leading to nitrogen leaching), soil erosion, reduction of crop diversity

- ◉ Rural space, through the loss and gain of cultivated lands, land speculation, land renunciation, and hydraulic amenities.

- ◉ Adaptation, organisms may become more or less competitive, as well as humans may develop urgency to develop more competitive organisms, such as flood resistant or salt resistant varieties of rice.

They are large uncertainties to uncover, particularly because there is lack of information on many specific local regions, and include the uncertainties on magnitude of climate change, the effects of technological changes on productivity, global food demands, and the numerous possibilities of adaptation.

Most agronomists believe that agricultural production will be mostly affected by the severity and pace of climate change, not so much by gradual trends in climate. If change is gradual, there may be enough time for biota adjustment. Rapid climate change, however, could harm agriculture in many countries, especially those that are already suffering from rather poor soil and climate conditions, because there is less time for optimum natural selection and adaption.

But much remains unknown about exactly how climate change may affect farming and food security, in part because the role of farmer behaviour is poorly captured by crop-climate models. For instance, Evan Fraser, a geographer at the University of Guelph in Ontario Canada, has conducted a number of studies that show that the socio-economic context of farming may play a huge role in determining whether a drought has a major, or an insignificant impact on crop production. In some cases, it seems that even minor droughts have big impacts on food security (such as what happened in Ethiopia in the early 1980s where a minor drought triggered a massive famine), versus cases where even relatively large weather-related problems were adapted to without much hardship. Evan Fraser combines socio-economic models along with climatic models to identify "vulnerability hotspots" One such study has identified US maize (corn) production as particularly vulnerable to climate change because it is expected to be exposed to worse droughts, but it does not have the socio-economic conditions that suggest farmers will adapt to these changing conditions. Other studies rely instead on projections of key agro-meteorological or agro-climate indices, such as growing season length, plant heat stress, or start of field operations, identified by land management stakeholders and that provide useful information on mechanisms driving climate change impact on agriculture.

9.6 PEST INSECTS AND CLIMATE CHANGE

Global warming could lead to an increase in pest insect populations, harming yields of staple crops like wheat, soybeans, and corn. While warmer temperatures create longer growing seasons, and faster growth rates for plants, it also increases the metabolic rate and number of breeding cycles of insect populations. Insects that previously had only two breeding cycles per year could gain an additional cycle if warm growing seasons extend, causing a population boom. Temperate places and higher latitudes are more likely to experience a dramatic change in insect populations.

The University of Illinois conducted studies to measure the effect of warmer temperatures on soybean plant growth and Japanese beetle populations. Warmer temperatures and elevated CO_2 levels were simulated for one field of soybeans, while the other was left as a control. These studies found that the soybeans with elevated CO_2 levels grew much faster and had higher yields, but attracted Japanese beetles at a significantly higher rate than the control field. The beetles in the field with increased CO_2 also laid more eggs on the soybean plants and had longer lifespans, indicating the possibility of a rapidly expanding population. DeLucia projected that if the project were to continue, the field with elevated CO_2 levels would eventually show lower yields than that of the control field.

The increased CO_2 levels deactivated three genes within the soybean plant that normally create chemical defenses against pest insects. One of these defenses is a protein that blocks digestion of the soy leaves in insects. Since this gene was deactivated, the beetles were able to digest a much higher amount of plant matter than the beetles in the control field. This led to the observed longer lifespans and higher egg-laying rates in the experimental field.

There are a few proposed solutions to the issue of expanding pest populations. One proposed solution is to increase the number of pesticides used on future crops. This has the benefit of being relatively cost effective and simple, but may be ineffective. Many pest insects have been building up an immunity to these pesticides. Another proposed solution is to utilize biological control agents. This includes things like planting rows of native vegetation in between rows of crops. This solution is beneficial in its overall environmental impact. Not only are more native plants getting planted, but pest insects are no longer building up an immunity to pesticides. However, planting additional native plants requires more room, which destroys additional acres of public land. The cost is also much higher than simply using pesticides.

9.6.1 Plant Diseases and Climate Change

Although research is limited, research has shown that climate change may alter the developmental stages of pathogens that can affect crops. The biggest consequence of climate change on the dispersal of pathogens is that the geographical distribution of hosts and pathogens could shift, which would result in more crop losses. This could affect competition and recovery from disturbances of plants. It has been predicted that the effect of climate change will add a level of complexity to figuring out how to maintain sustainable agriculture.

9.6.2 Observed Impacts

Effects of regional climate change on agriculture have been limited. Changes in crop phenology provide important evidence of the response to recent regional climate change. Phenology is the study of natural phenomena that recur periodically, and how these phenomena relate to climate and seasonal changes. A significant advance in phenology has been observed for agriculture and forestry in large parts of the Northern Hemisphere.

Droughts have been occurring more frequently because of global warming and they are expected to become more frequent and intense in Africa, southern Europe, the Middle East, most of the Americas, Australia, and Southeast Asia. Their impacts are aggravated because of increased water demand, population growth, urban expansion, and environmental protection efforts in many areas. Droughts result in crop failures and the loss of pasture grazing land for livestock.

Examples

As of the decade starting in 2010, many hot countries have thriving agricultural sectors.

Jalgaon district, India, has an average temperature which ranges from 20.2 °C in December to 29.8 °C in May, and an average precipitation of 750 mm/year. It produces bananas at a rate that would make it the world's seventh-largest banana producer if it were a country.

During the period 1990-2012, Nigeria had an average temperature which ranged from a low of 24.9 °C in January to a high of 30.4 °C in April. According to the Food and Agriculture Organization of the United Nations (FAO), Nigeria is by far the world's largest producer of yams, producing over 38 million tonnes in 2012. The second through

8th largest yam producers were all nearby African countries, with the largest non-African producer, Papua New Guinea, producing less than 1% of Nigerian production.

In 2013, according to the FAO, Brazil and India were by far the world's leading producers of sugarcane, with a combined production of over 1 billion tonnes, or over half of worldwide production.

In the summer of 2018, heat waves probably linked to climate change cause much lower than average yield in many parts of the world, especially in Europe. Depending on conditions during August, more crop failures could rise global food prices. losses are compared to those of 1945, the worst harvest in memory. last year was the third time in four years that global wheat, rice and maize production failed to meet demand, forcing governments and food companies to release stocks from storage. India last week released 50% of its food stocks. Lester Brown, the head of Worldwatch, an independent research organisation, predicted thatfood prices will rise in the next few months.

Overall food shortages are not expected this year. But, for prevent hunger, instability, new waves of Climate refugees international help to countries who will luck the money to buy enough food and stopping conflicts will be needed.

9.7 PROJECTIONS OF IMPACTS

As part of the IPCC's Fourth Assessment Report, Schneider et al. (2007) projected the potential future effects of climate change on agriculture. With low to medium confidence, they concluded that for about a 1 to 3 °C global mean temperature increase (by 2100, relative to the 1990–2000 average level) there would be productivity decreases for some cereals in low latitudes, and productivity increases in high latitudes. In the IPCC Fourth Assessment Report, "low confidence" means that a particular finding has about a 2 out of 10 chance of being correct, based on expert judgement. "Medium confidence" has about a 5 out of 10 chance of being correct. Over the same time period, with medium confidence, global production potential was projected to:

⦿ Increase up to around 3 °C,

⦿ Very likely decrease above about 3 °C.

Most of the studies on global agriculture assessed by Schneider et al. (2007) had not incorporated a number of critical factors, including changes in extreme events, or the spread of pests and diseases. Studies had also not considered the development of specific practices or technologies to aid adaptation to climate change.

The US National Research Council assessed the literature on the effects of climate change on crop yields. US NRC (2011) stressed the uncertainties in their projections of changes in crop yields.

Writing in the journal Nature Climate Change, Matthew Smith and Samuel Myers (2018) estimated that food crops could see a reduction of protein, iron and zinc content in common food crops of 3 to 17%. This is the projected result of food grown under the expected atmospheric carbon-dioxide levels of 2050. Using data from the UN Food and Agriculture Organization as well as other public sources, the authors analyzed 225 different staple foods, such as wheat, rice, maize, vegetables, roots and fruits.

Their central estimates of changes in crop yields are shown above. Actual changes in yields may be above or below these central estimates. US NRC (2011) also provided an estimated the "likely" range of changes in yields. "Likely" means a greater than 67% chance of being correct, based on expert judgement. The likely ranges are summarized in the image descriptions of the two graphs.

9.7.1 Food Security

The IPCC Fourth Assessment Report also describes the impact of climate change on food security. Projections suggested that there could be large decreases in hunger globally by 2080, compared to the (then-current) 2006 level. Reductions in hunger were driven by projected social and economic development. For reference, the Food and Agriculture Organization has estimated that in 2006, the number of people undernourished globally was 820 million. Three scenarios without climate change (SRES A1, B1, B2) projected 100-130 million undernourished by the year 2080, while another scenario without climate change (SRES A2) projected 770 million undernourished. Based on an expert assessment of all of the evidence, these projections were thought to have about a 5-in-10 chance of being correct.

The same set of greenhouse gas and socio-economic scenarios were also used in projections that included the effects of climate change. Including climate change, three scenarios (SRES A1, B1, B2) projected 100-380 million undernourished by the year 2080, while another scenario with climate change (SRES A2) projected 740-1,300 million undernourished. These projections were thought to have between a 2-in-10 and 5-in-10 chance of being correct.

Projections also suggested regional changes in the global distribution of hunger. By 2080, sub-Saharan Africa may overtake Asia as the world's most food-insecure region. This is mainly due to projected social and economic changes, rather than climate change.

In South America, a phenomenon known as the El Nino Oscillation Cycle, between floods and drought on the Pacific Coast has made as much as a 35% difference in Global yields of wheat and grain.

Looking at the four key components of food security we can see the impact climate change has had. Food "Access to food is largely a matter of household and individual-level income and of capabilities and rights". Access has been affected by the thousands of crops being destroyed, how communities are dealing with climate shocks and adapting to climate change. Prices on food will rise due to the shortage of food production due to conditions not being favourable for crop production. Utilization is affected by floods and drought where water resources are contaminated, and the changing temperatures create vicious stages and phases of disease. Availability is affected by the contamination of the crops, as there will be no food process for the products of these crops as a result. Stability is affected through price ranges and future prices as some food sources are becoming scarce due to climate change, so prices will rise.

9.7.2 Individual Studies

Cline (2008) looked at how climate change might affect agricultural productivity in the 2080s. His study assumes that no efforts are made to reduce anthropogenic greenhouse gas emissions, leading to global warming of 3.3 °C above the pre-industrial level. He concluded that global agricultural productivity could be negatively affected by climate change, with the worst effects in developing countries (see graph opposite).

Lobell et al. (2008a) assessed how climate change might affect 12 food-insecure regions in 2030. The purpose of their analysis was to assess where adaptation measures to climate change should be prioritized.

They found that without sufficient adaptation measures, South Asia and South Africa would likely suffer negative impacts on several crops which are important to large food insecure human populations.

Battisti and Naylor (2009) looked at how increased seasonal temperatures might affect agricultural productivity. Projections by the IPCC suggest that with climate change, high seasonal temperatures will become widespread, with the likelihood of extreme temperatures increasing through the second half of the 21st century. Battisti and Naylor (2009) concluded that such changes could have very serious effects on agriculture, particularly in the tropics. They suggest that major, near-term, investments in adaptation measures could reduce these risks.

"Climate change merely increases the urgency of reforming trade policies to ensure that global food security needs are met" said C. Bellmann, ICTSD Programmes Director. A 2009 ICTSD-IPC study by Jodie Keane suggests that climate change could cause farm output in sub-Saharan Africa to decrease by 12% by 2080 - although in some African countries this figure could be as much as 60%, with agricultural exports declining by up to one fifth in others. Adapting to climate change could cost the agriculture sector $14bn globally a year, the study finds.

9.8 COMPARISON OF TEMPERATE AND TROPICAL AGRI-CULTURE

The tropics are defined as the geographical area lying between 23.5deg. N and 23.5deg.S latitude, while the temperate regions are found above these parallels. Climatologically, the tropics are characterized by high year-round temperatures and weather is controlled by equatorial and tropical air masses. Tropical precipitation is primarily convective. In the more humid tropical regions, annual rainfall is often above 2000 mm and falls in almost all months of the year. In the drier tropics, rainfall can fall below 50 mm, and be very seasonal. The remainder of the region lies between these precipitation regimes, with distinct wet and dry seasons. Agriculture is frequently limited by the seasonality and magnitude of moisture availability.

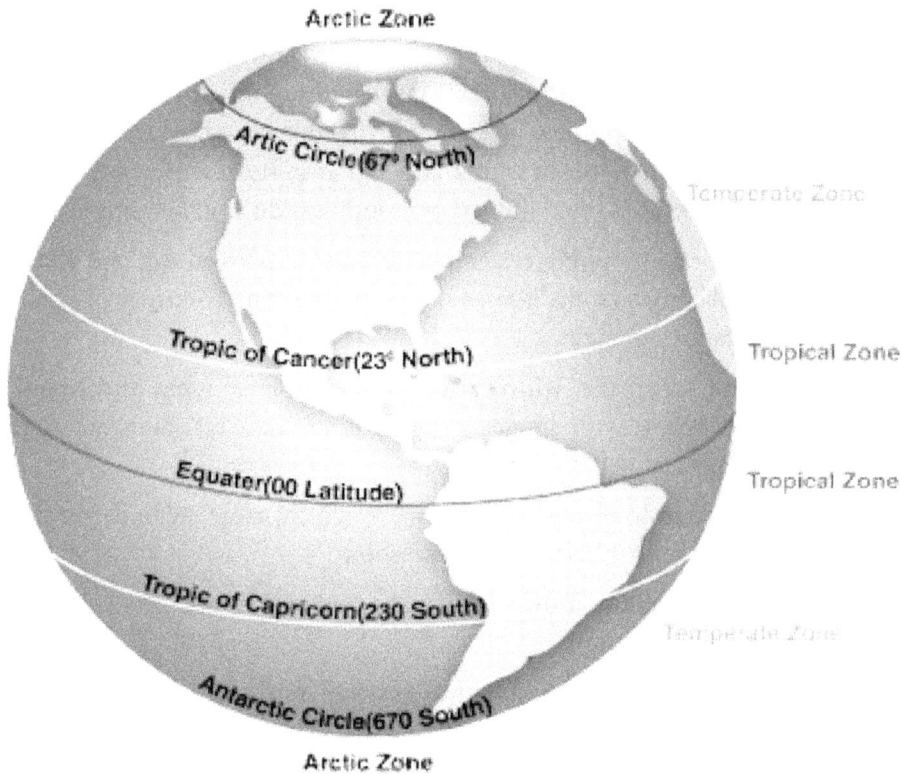

Fig. 4. Comparison of temperate and tropical agriculture.

In the mid-latitude temperate zone, weather is controlled by both tropical and polar air masses. Precipitation here occurs along fronts within cyclonic storms. The temperate region also has many different climate regions with warmer and cooler temperatures and seasonal rainfall. Temperate agriculture is often characterized a predominantly limited by seasonally cooler temperatures.

Reported experiments have shown that even though yield per day is often higher in the tropics, total crop growth season is shorter. Leaf area expansion and phasic development are faster in the tropics because of higher temperatures during vegetative growth. Nevertheless, crop yields are consistently found to be higher in temperate regions than in the tropics Numerousfactors contribute to this result. Soils in the humid tropics tend to be highly leached of nutrients and are therefore unproductive because of high temperatures, intense rainfall, and erosion. Soils in the drier tropics are often hampered by accumulations of salt and lack of water. Temperate soils are generally viewed as more favorable to agriculture than tropical soils because of higher nutrient levels. However, there are exceptions in both regions, with high productive volcanic and fluvial soils found in the tropics, and poorly developed and infertile soils in temperate regions.

Agricultural production is also severely limited in many humid tropical regions by the wide range of weeds, pests, and diseases that flourish in consistently warm and moist climates. The growth of some crops and varieties, which require long hours of daylight to reach maturity, is also limited by the invariable day lengths of the tropics. Solar radiation, which is critical to plant growth, and whose intensity is controlled by the angle of the sun, daylength, and cloudiness, is lower in winter and higher in summer in temperate zones. In the tropics, solar radiation is often limited by cloudiness during the rainy seasons.

Agricultural crops and cropping systems have been developed for, and adapted to, these varied regimes of climate, soil, diseases and pests. The main commercial agricultural crops and their adaptations include:

(a) Cassava and sugarcane, which only grow in tropical areas and have a crop duration of one year or longer. Cassava is drought resistant, but sugarcane requires irrigation in dry areas.

(b) Sorghum, groundnut, and sweet potato, which grow in both tropical and subtropical regions in relatively dry seasons.

(c) Rice, which is mainly grown in tropical and subtropical zones in the rainy season or with irrigation.

(d) Maize and field beans grow in both zones, preferably with seasons with enough rain

(e) Wheat, soybean, and potato are crops of the subtropical and temperate zones and grow in the tropics at high (cooler) elevations.

(f) Sugarbeet is grown only in the temperate zone.

A number of "luxury" agricultural crops, especially fruit (bananas, pineapples), stimulants (coffee, tea) and spices grow only, or best, in the tropics. Tropical regions are also important in providing winter season produce for temperate zones.

In temperate agriculture, plant breeding and fertilizer use produced dramatic yield increases for many crops early in the twentieth century. Similar increases occurred more recently in tropical regions for crops such as wheat, maize and rice, which benefited from the technological package or improved seeds, fertilizer, mechanization and pesticides known as the Green Revolution.

In both temperate and tropical regions, irrigation has been developed in areas where dry seasons exist and adequate water can be reserved from other seasons or brought in from adjacent regions. Irrigation is an important buffer against climate variability and climate change. About 20% of the world's cropland is irrigated, mostly in Asia, producing about 40% of the annual crop production.

Differences in farming systems, technology and economics also contribute to the yield differences in temperate and tropical regions. Agriculture in temperate regions is characterized by high levels of inputs (quality seed stock, fertilizer, herbicides and pesticides), and a high degree of mechanization and capitalization. However, there are wide variations in the use of technology, European agriculture being particularly intensive. In tropical regions, many farmers cannot afford inputs, and governments cannot afford to subsidize them. In some parts of the tropics, traditional technologies, such as multiple cropping and terracing, act to buffer the system against climate variability, conserve soil fertility, and increase yields.

In some senses, the tropics are more dependent on agriculture, and therefore more vulnerable to climatic change, than the temperate regions. As much as 75% of the world's population live in the tropics, and two thirds of these people are reliant on agriculture for their livelihoods. With low levels of technology, land degradation, unequal land distribution, and rapid population growth, many tropical regions are near or exceeding their capacity to feed themselves.

Indeed, some authors have argued that the unequal social structures and international position of many tropical countries also increase vulnerability to climate change. Unequal land tenure, high numbers of landless rural dwellers, low incomes and high national debts exacerbate the negative impacts of climate variability, as some people have no extra land, job, savings or government assistance to see them through droughts or other climatic extremes. When the economic system is oriented towards export rather than subsistence agriculture, climatic change (as well as low export prices) may threaten the whole national economy and food system. Those regions that cannot feed their populations depend on cereal imports from the major cereal exporters such as the USA, France, Canada, Australia, Argentina, and Thailand. All except Thailand would be defined as temperate agricultural producers.

9.9 THE INFLUENCE OF CLIMATE CHANGE ON CROP PRODUCTION

At the basis of any understanding of climate impacts on agriculture lies the biophysical sciences. The rates of most biophysical processes are highly dependent on climate variables such as radiation, temperature, and moisture, that vary regionally. For example, rates of plant photosynthesis depend on the amount of photosynthetically active radiation and levels of atmospheric carbon dioxide (CO_2). Temperature is an important determinant of the rate at which a plant progresses through various phenological stages towards maturity. The accumulation of biomass is constrained by the availability of moisture and nutrients to a growing plant.

Numerous studies have examined the impacts of past climatic variations on agriculture using case studies, statistical analyses and simulation models. Such studies have clearly demonstrated the sensitivity of both temperate and tropical agricultural systems and nations to climatic variations and changes. In the temperate regions, the impacts of climate variability, particularly drought, on yields of grains in North America and the Soviet Union have been of particular concern because of their effects on world food security. In the tropics, drought impacts on agriculture and resulting food shortages have been widely studied, especially when associated with the failure of the monsoon in Asia or the rains in Sudano-Sahelian Africa. In the temperate regions, climatic variations are associated with economic disruptions; in the tropics, droughts bring famine and widespread social unrest.

9.9.1 The Biophysical Impact of Climate Change Associated with Global Warming

It is frequently assumed that global change will bring higher temperatures, altered precipitation, and higher levels of atmospheric CO_2. What might these changes mean for the biophysical response of agricultural crops?

Interactions with thermal regimes. Higher temperatures in general hasten plant maturity in annual species, thus shortening the growth stages during which pods, seeds, grains or bolls can absorb photosynthetic products. This is one reason yield are lower in the tropics. Because crop yield depends on both the rate of carbohydrate accumulation and the duration of the filling periods, the economic yields of both temperate and tropical crops grown in a warmer and CO_2-enriched environment may not rise substantially above present levels, despite increases in net photosynthesis.

Because temperature and tropical regions differ in both current temperature and the temperature rise predicted for climate change, the relative magnitudes of combined CO_2 and temperature effects will likely be different in the different regions. In the mid-latitudes, higher temperatures may shift biological process rates toward optima, and beneficial effects are likely to ensue. Increases in temperature will also lengthen the frost-free season in temperate regions, allowing for longer duration crop varieties to be grown and offering the possibility of growing successive crops (moisture conditions permitting). In tropical locations where increased temperatures may move beyond optima, negative consequences may dominate.

Both the mean and extreme temperatures that crops experience during the growing season will change in both temperate and tropical areas. Extreme temperatures are important because many crops have critical thresholds both above and below which crops are damaged. Prolonged hot spells can be especially damaging. Critical stages for high temperature injury include seedling emergence in most crops, silking and tasseling in corn, grain filling in wheat, and flowering in soybeans. In general, higher temperatures should decrease cold damage and increase heat damage. Agro-climatic zones are expected to shift poleward as lengthening and warming growing seasons allow new or enhanced crop production (soil resources permitting).

Changes in hydrological regimes. The hydrological regimes in which crops grow will surely change with global warming. While all GCMs predict increases in mean global precipitation (because a warmer atmosphere can hold more water vapor), decreases are forecast in some regions and increases are not uniformly distributed. The crop water regime may further be affected by changes in seasonal precipitation, within-season pattern of precipitation, and interannual variation of precipitation. Increased convective rainfall is predicted to occur, particularly in the tropics, caused by stronger convection cells and more moisture in the air.

Too much precipitation can cause disease infestation in crops, while too little can be detrimental to crop yields, especially if dry periods occur during critical development stages. For example, moisture stress during the flowering, pollination, and grain-filling stages is especially harmful to maize, soybean, wheat and sorghum.

The amount and availability of water stored in the soil, a crucial input to crop growth, will be affected by changes in both the precipitation and seasonal and annual evapotranspiration regimes. Some GCMs predict mid-continental drying the Northern Hemisphere and other GCM predictions have been interpreted to suggest that the rise in potential evapotranspiration will exceed that of rainfall resulting in drier regimes throughout the tropics and low to mid-latitudes. Because the soil moisture processes are represented so crudely in the current GCMs, however, it is difficult to associate much certainty with these projections.

Global climate change is likely to exacerbate the demand for irrigation water. Higher temperatures, increased evaporation, and yield decreases contribute to this projection. However, supply of needed irrigation water under climate change in uncertain. Where water supplies are diminishing, such as the Ogallala Aquifer in the United States, extra demand might require that some land be withdrawn from irrigation.

Physiological effects of CO_2. The study of agricultural impacts of trace gas induced climate change is complicated by the fact that increasing atmospheric CO_2 has other effects on crop plants besides its alteration of their climate regime. These are often called "fertilizing" effects, because of their perceived beneficial physiological nature. Specifically, most plants growing in enhanced CO_2 exhibit increased rates of net photosynthesis. The higher photosynthesis rates are then manifested in higher leaf area, dry matter production, and yield for many crops. In several cases, high CO_2 has contributed to upward shifts in temperature optima for photosynthesis (Jurik et al., 1984) and to enhanced growth with higher temperatures; other studies, however, have not shown such benefits.

CO_2 enrichment also tends to close plant stomates, and by doing so, reduces transpiration per unit leaf area while still enhancing photosynthesis. The stomatal conductances of 18 agricultural species have been observed to decrease markedly (by 36%, on average) in an atmosphere enriched by doubled CO_2. However, crop transpiration per ground area may not be reduced commensurately, because decreases in individual leaf conductance tend to be offset by increases in crop leaf area. In any case, higher CO_2 often improves water-use efficiency, defined as the ratio between crop biomass accumulation or yield and the amount of water used in evapotranspiration. Increases in photosynthesis and resistance with higher CO_2 have been shown to occur at less than optimal levels of other environmental variables, such as light, water, and some of the mineral nutrients.

Temperate crops may benefit more from increasing CO_2 than tropical crops. In crop species with the C_3 pathway characteristic of non-tropical plants (e.g., wheat, soybean, cotton) CO_2 enrichment has been shown to decrease photorespiration, the rapid oxidation of recently formed sugars in the light, a process which lowers the efficiency of overall photosynthesis. C4 crops, which are particularly characteristic of tropical and warm arid regions (e.g., maize, sorghum, and millet), are more efficient photosynthetically under current CO_2 levels than C_3 plants (because they fix CO_2 into malate in their mesophyll cells before delivering it to the RuBP enzyme in the bundle-sheath cells). Because of this CO_2-concentrating and photorespiration-avoiding mechanism, experimental data show that C4 plants are less responsive to CO^2 enrichment.

The physiological effects of high levels of atmospheric CO_2 described above have been observed under controlled experimental conditions. In the open field, however, their magnitude and significance are still largely untested, and their importance relative to the predicted large-scale climatic effects uncertain. Greenhouse and field-chamber environments tend to be much smaller, less variable, and more protected from wind than field conditions. Furthermore, physiological feedback mechanisms such as starch accumulation or lack of sink (that is, growing, storing, or metabolizing tissue) for the products of photosynthesis may limit the extent to which the "fertilizing" CO_2 effects may be realized. Finally, if trace gas emissions continue to grow unchecked, their climate warming effect is projected to continue even up to 2000 ppm, but the beneficial boost to photosynthesis appears to level off at about 400 ppm for C_4 crops and about 800 ppm for C_3 crops.

Soils. Climate change will also have an impact on the soil, a vital element in agricultural ecosystems. Higher air temperatures will cause higher soil temperatures, which should generally increase solution chemical reaction rates and diffusion-controlled reactions. Solubilities of solid and gaseous components may either increase or decrease, but the consequences of these changes may take many years to become significant. Furthermore, higher temperatures will accelerate the decay of soil organic matter, resulting in release of CO_2 to the atmosphere and decrease in carbon/nitrogen ratios, although these two effects should be offset somewhat by the greater root biomass and crop residues resulting from plant responses to higher CO_2.

In temperate countries where crops are already heavily fertilized, there will probably be no major changes in fertilization practices, but alterations in timing and method (e.g., careful adjustment of side-dress applications of nitrogen during vegetative crop growth) are expected with changes in temperature and precipitation regimes. In tropical countries, where fertilization level is not always adequate, the need for fertilization will probably increase.

Sea level rise, another predicted effect of global warming, will caused increased flooding, salt-water intrusion, and rising water tables in agricultural soils located near coastlines. This is particularly crucial in tropical countries such as Bangladesh, with large agricultural regions and high rural population located near current sea level.

Pests are organisms that affect agricultural plants and animals in ways considered unfavorable. They include weeds, and certain insects, arthropods, nematodes, bacteria, fungi, and viruses. Because climate variables (especially temperature, wind and humidity) control the geographic distribution of pests, climate change is likely to alter their ranges. Insects may extend their ranges where warmer winter temperatures allow their over-wintering survival and increase the possible number of generations per season. Pests and diseases from low latitude regions, where they are much more prevalent may be introduced at higher latitudes. As a consequence of pest increase, there may be a substantial rise in the use of agricultural chemicals in both temperate and tropical regions to control them.

10

ORGANIC FARMING MANAGEMENT

10.1 OBJECTIVES

After studying this chapter you will be able to understand:

- Organic Farming Techniques
- Disease and Pest Resistance
- Scope of Organic Farming.
- Organic Agricultural Movements.

10.2 INTRODUCTION

Organic farming is a system that is dependant on skilled management – use of legumes, recycling of nutrients, clean grazing for intestinal worm control, encouraging pest predators, crop and animal selection and breeding and careful timing of cultivations to avoid weeds – rather than the use of inputs of fertilisers, herbicides, pesticides and veterinary medicines.

Diversity is a characteristic of organic farms and some are taking this another step by adopting agroforestry methods.

The results are impressive in terms of delivering the broad spectrum of public goods, which are required of all farms in the new Defra Environmental Land Management Scheme, along with quality food and a profitable business.

Organic management is not necessarily more difficult, but it certainly requires a high standard of management and the use of many techniques that are not usually used in conventional farming. In particular, producing intensive and field scale organic vegetables demands precise use of some specific and sometimes very advanced weed control machinery, such as the camera guided inter row and in-row hoes.

Soil management is central to organic farming, understanding your soil through digging, observation and analysis is an essential start to stimulating soil biological activity and ensuring that there are no underlying nutrient deficiencies which might need the use of mineral fertilisers.

Marketing, as with any business, is critical. While some farms will focus on wholesale markets, such as the substantially undersupplied grain market, others will want to supply local markets and maybe develop their own processing and retail outlets. "Organic" provides the most effective marketing label for independently certified, legally defined quality food.

10.3 ORGANIC FARMING TECHNIQUES

Combine scientific knowledge of ecology and modern technology with traditional farming practices based on naturally occurring biological processes. Organic farming methods are studied in the field of agroecology. While conventional agriculture uses synthetic pesticides and water-soluble synthetically purified fertilizers, organic farmers are restricted by regulations to using natural pesticides and fertilizers. The principal methods of organic farming includecrop rotation, green manures and ompost, biological pest control, and mechanical cultivation.

These measures use the natural environment to enhance agricultural productivity: legumes are planted to fix nitrogen into the soil, natural insect predators are encouraged, crops are rotated to confuse pests and renew soil, and natural materials such as potassium bicarbonate and mulches are used to control disease and weeds. Organic farmers are careful in their selection of plant breeds, and organic researchers produce hardier plants through plant breeding rather than genetic engineering.

10.3.1 Crop Diversity

Crop diversity is a distinctive characteristic of organic farming. Conventional farming focuses on mass production of one crop in one location, a practice called monoculture. This makes apparent economic sense: the larger the growing area, the lower the per unit cost of fertilizer, pesticides and specialized machinery for a single plant species. The science of agroecology has revealed the benefits of polyculture (multiple crops in the same space), which is often employed in organic farming. Planting a variety of vegetable crops supports a wider range of beneficial insects, soil microorganisms, and other factors that add up to overall farm health, but managing the balance requires expertise and close attention.

10.3.2 Farm Size

Farm size in great measure determines the general approach and specific tools and methods. Today, major food corporations are involved in all aspects of organic production on a large scale. However, organic farming originated as a small-scale enterprise, with operations from under 1-acre (4,000 m2) to under 100 acres (0.40 km2). The mixed vegetable organic market garden is often associated with fresh, locally grown produce, farmers' markets and the like, and this type of farm is often under 10 acres (40,000 m2). Farming at this scale is generally labor-intensive, involving more manual labor and less mechanization.

The type of crop also determines size: organic grain farms often involve much larger area. Larger organic farms tend to use methods and equipment similar to conventional farms, centered around the tractor.

10.3.3 Plant Nutrition

The central farming activity of fertilization illustrates the differences. Organic farming relies heavily on the natural breakdown of organic matter, using techniques like green manure and composting, to replace nutrients taken from the soil by previous crops. This biological process, driven by microorganisms such as mycorrhiza, allows the natural production of nutrients in the soil throughout the growing season, and has been referred to as feeding the soil to feed the plant. In chemical farming, individual nutrients, like nitrogen, are synthesized in a more or less pure form that plants can use immediately, and applied on a man-made schedule. Each nutrient is defined and addressed separately. Problems that may arise from one action (e.g. too much nitrogen left in the soil) are usually addressed with additional, corrective products and procedures (e.g. using water to wash excess nitrogen out of the soil).

Organic farming uses a variety of methods to improve soil fertility, including crop rotation, cover cropping, reduced tillage, and application of compost. By reducing tillage, soil is not inverted and exposed to air; less carbon is lost to the atmosphere resulting in more soil organic carbon. This has an added benefit of carbon sequestration which can reduce green house gases and aid in reversing climate change.

10.3.4 Pest Control

Different approaches to pest control are equally notable. In chemical farming, a specific insecticide may be applied to quickly kill off a particular insect pest (animal). Chemical controls can dramatically reduce pest populations for the short term, yet by unavoidably killing (or starving) natural predator insects and animals, cause an ultimate increase in the pest population. Repeated use of insecticides and herbicides and other pesticides also encourages rapid natural selection of resistant insects, plants and other organisms, necessitating increased use, or requiring new, more powerful controls.

In contrast, organic farming tends to tolerate some pest populations while taking a longer-term approach. Organic pest control involves the cumulative effect of many techniques, including:

- Allowing for an acceptable level of pest damage;
- Encouraging predatory beneficial insects to control pests;

⦿ Encouraging beneficial microorganisms and insects; this by serving them nursery plants and/or an alternative habitat, usually in a form of a shelterbelt, hedgerow, or beetle bank

⦿ Careful crop selection, choosing disease-resistant varieties

⦿ Planting companion crops that discourage or divert pests;

⦿ Using row covers to protect crops during pest migration periods;

⦿ Using pest regulating plants and biologic pesticides and herbicides

⦿ Using no-till farming, and no-till farming techniques as false seedbeds

⦿ Rotating crops to different locations from year to year to interrupt pest reproduction cycles;

⦿ Using insect traps to monitor and control insect populations.

Each of these techniques also provides other benefits—soil protection and improvement, fertilization, pollination, water conservation, season extension, etc.—and these benefits are both complementary and cumulative in overall effect on farm health. Effective organic pest control requires a thorough understanding of pest life cycles and interactions.

10.4 LIVESTOCK

Raising livestock and poultry, for meat, dairy and eggs, is another traditional, farming activity that complements growing. Organic farms attempt to provide animals with "natural" living conditions and feed. While the USDA does not require any animal welfare requirements be met for a product to be marked as organic, this is a variance from older organic farming practices. Ample, free-ranging outdoor access, for grazing and exercise, is a distinctive feature, and crowding is avoided. Feed is also organically grown, and drugs, including antibiotics, are not ordinarily used (and are prohibited under organic regulatory regimes). Animal health and food quality are thus pursued in a holistic "fresh air, exercise, and good food" approach.

Also, horses and cattle used to be a basic farm feature that provided labor, for hauling and plowing, fertility, through recycling of manure, and fuel, in the form of food for farmers and other animals. While today, small growing operations often do not include livestock, domesticated animals are a desirable part of the organic farming equation, especially for true sustainability, the ability of a farm to function as a self-renewing unit.

10.4.1 Organic Farming Systems

There are several organic farming systems. Biodynamic farming is a comprehensive approach, with its own international governing body. The Do Nothing Farming method focuses on a minimum of mechanical cultivation and labor for grain crops. French intensive and biointensive, methods are well-suited to organic principles. Other techniques are permaculture and no-till farming. Finally, newcomers as the Agro-ecologic system focus on a blend of a more large-scale approach with imbedded natural/organic farming techniques. A farm may choose to adopt a particular method, or a mix of techniques.

While fundamentally different, large-scale agriculture and organic farming are not entirely mutually exclusive. For example, Integrated Pest Management is a multifaceted strategy that can include synthetic pesticides as a last resort—both organic and conventional farms use IPM systems for pest control.

10.4.2 Capacity Building

Organic agriculture can contribute to ecologically sustainable, socio-economic development, especially in poorer countries. The application of organic principles enables employment of local resources (e.g., local seed varieties, manure, etc.) and therefore cost-effectiveness. Local and international markets for organic products show tremendous growth prospects and offer creative producers and exporters excellent opportunities to improve their income and living conditions.

Organic agriculture is knowledge intensive. Globally, capacity building efforts are underway, including localized training material, to limited effect. As of 2007, the International Federation of Organic Agriculture Movements hosted more than 170 free manuals and 75 training opportunities online.

10.5 CONTROVERSY

Norman Borlaug, father of the "Green Revolution", Nobel Peace Prize laureate, Prof A. Trewavas and other critics contested the notion that organic agricultural systems are more friendly to the environment and more sustainable than conventional farming systems. Borlaug asserts that organic farming practices can at most feed 4 billion people, after expanding cropland dramatically and destroying ecosystems in the process. The Danish Environmental Protection Agency estimated that phasing out all pesticides would result in an overall yield reduction of about 25%. Environmental and health effects were assumed but hard to assess.

In contrast, the UN Environmental Programme concluded that organic methods greatly increase yields in Africa. A review of over two hundred crop comparisons argued that organic farming could produce enough food to sustain the current human population and that the difference in yields between organic and non-organic methods were small, with non-organic methods yielding slightly more in developed areas and organic methods yielding slightly more in developing areas.

That analysis has been criticised by Alex Avery of the Hudson Institute, who contends that the review claimed many non-organic studies to be organic, misreported organic yields, made false comparisons between yields of organic and non-organic studies which were not comparable, counted high organic yields several times by citing different papers which referenced the same data, and gave equal weight to studies from sources which were not impartial.[100] The Center for Disease Control repudiated a claim by Avery's father, Dennis Avery (also at Hudson) that the risk of E. coli infection was eight times higher when eating organic food. (Avery had cited CDC as a source.) Avery had included problems stemming from non-organic unpasteurized juice in his calculations. However, the 2011 E. coli O104:H4 outbreak, which has caused over 3,900 cases and 52 deaths was traced by epidemiologists to an organic farm in Bienenbüttel, Germany.

Urs Niggli, director of the FiBL Institute contends that there is a global campaign against organic farming that mostly derives from Avery's book The truth about organic farming

10.6 ADVANTAGES AND DISADVANTAGES OF ORGANIC FARMING

Despite the good things about organic farming why do most farmers still operate by industrialized agriculture?

Here we explore the pros and cons organic farming presents for consumers and producers, as well as examining the environmental effects of organic farming.

10.6.1 Good Things About Organic Farming Consumer Benefits

Nutrition

The nutritional value of food is largely a function of its vitamin and mineral content. In this regard, organically grown food is dramatically superior in mineral content to that grown by modern conventional methods. farming

Because it fosters the life of the soil organic farming reaps the benefits soil life offers in greatly facilitated plant access to soil nutrients.

Healthy plants mean healthy people, and such better nourished plants provide better nourishment to people and animals alike.

10.6.2 Poison-Free Advantages and Disadvantages of Organic Farming

A major benefit to consumers of organic food is that it is free of contamination with health harming chemicals such as pesticides, fungicides and herbicides.

As you would expect of populations fed on chemically grown foods, there has been a profound upward trend in the incidence of diseases associated with exposure to toxic chemicals in industrialized societies.

Take cancer for example. Representative data on the number of new cancer cases in New South Wales, Australia has been collected by the New South Wales Central Cancer Registry.

Adjusted to take account of our aging population, their graph (above) shows that between 1972 and 2004 the incidence of new cancer cases per year (average for both sexes) has risen from 323 to 488 per 100,000 people. This is an increase of over 50% in just 32 years.

10.7 GM CROPS

Organic growers do not use genetically modified or engineered food crops, some of which are engineered to tolerate herbicides (e.g. "Roundup Ready Canola") or resist pests (e.g. Bollworm resistant cotton). Conventional growers, on the other hand, are free to "take advantage" of GM crops.

According to a report from the Directorate-General for Agriculture of the European Commission, productivity gains attributed to GM crops are usually negligible when growing conditions, farmer experience and soil types are factored in, and are often in fact negative. The main advantage farmers using such crops gain is convenience only.

There are worrying indications that GM crops may be associated with harm to both human health and the environment. The main concern is that once they are released it is nigh impossible to "un-release" them.

10.7.1 Time

Indeed, organic farming requires greater interaction between a farmer and his crop for observation, timely intervention and weed control for instance. It is inherently more labor intensive than chemical/mechanical agriculture so that, naturally a single farmer can produce more crop using industrial methods than he or she could by solely organic methods.

10.7.2 Skill

It requires considerably more skill to farm organically. However, because professional farming of any sort naturally imparts a close and observant relationship to living things, the best organic farmers are converted agrichemical farmers.

Organic farmers do not have some convenient chemical fix on the shelf for every problem they encounter. They have to engage careful observation and greater understanding in order to know how to tweak their farming system to correct the cause of the problem rather than simply putting a plaster over its effect.

This is a bigger issue during the conversion period from conventional to wholly organic when both the learning curve and transition related problems are peaking (it takes time to build a healthy farm ecosystem that copes well without synthetic crutches). Organic farmers I have interviewed report that their most valuable remedies and advice come from other organic farmers.

10.8 ENVIRONMENTAL EFFECTS OF ORGANIC FARMING

10.8.1 Climate Friendly

The synthetic inputs upon which conventional agriculture is so dependent are energy expensive to mine and manufacture. Today the embodied energy of industrial agriculture uses up 9 calories for every 1 calorie of food that it produces!

Organic agriculture with its low input needs of naturally derived substances produces less greenhouse gas emissions and is considerably more climate friendly.

10.8.2 Ecologically Friendly

It Doesn't Use Soluble Fertilizers

Though rarely acknowledged, the chief source of the annual algae blooms that plague Perth's major river (the Swan) is conventional agriculture.

Farmers pour tons of phosphate and nitrogenous fertilizer on their cropping lands every year. Because it is soluble, much of this fertilizer is either washed off the soil surface and into waterways (especially phosphates) or leaches through the soil profile beyond the reach of plants and finds its way less directly into waterways (especially nitrates). Nitrate contamination of groundwater (indicated by >10 mg/L nitrate) in Australia is widespread in every state and territory, occurring over regional and local scales. In many areas, the concentration is greater than the recently revised Australian Drinking Water Guidelines level of 50 mg/L nitrate (as nitrate), resulting in groundwater that is unfit for drinking. In some of the more contaminated areas, the concentration is in excess of 100 mg/L.

With fresh water reserves under increasing pressure from climate change this is a grave situation for humanity.

The soluble nutrient pollutants that contaminate surface waters fuel the overgrowth of algae. What is not used up by algae in fresh waterways, spews out into the ocean where it supports the growth of algae on sea plants and coral reef systems. This blocks access to sunlight, causing whatever it smothers to die.

Eighty percent of the seagrass meadows in Perth's Cockburn Sound – an important nursery habitat for wild fish stocks - have been decimated due to this process which is called "eutrophication".

It Doesn't Use Pesticides or Herbicides

Another pollution disaster caused by agrichemical use is the contamination of groundwater reserves with poisonous nasties, particularly (in Australia) Atrazine and Simazine, but also Dieldrin, Chlorpyriphos, Amitrol, Metolachlor, Trifluraline and Diuron Dieldrin, Lindane, and Alachlor.

While systematic monitoring of pesticide contamination of groundwater in Australia is limited, available tests have detected pesticides in at least 20% of samples, indicating significant contamination.

Groundwater studies in the US have found similarly significant contamination. In Carolina, for example, over 27% of wells sampled in 1997 were found to be contaminated with pesticides predominantly from routine agricultural usage.

There is no economically viable method to clean up widespread contamination. Pesticide contamination poses a serious, unreasonable public health threat to current and future ground water users. and disadvantages organic farming

Synthetic agrichemicals (and most plastics widely used in our society) are derived from oil, and thus a source of endocrine-disrupting chemicals (especially xenoestrogens) in the environment. Distorted sex organ development and function in alligators has been related to a major pesticide spill into a lake in Florida, U.S.A.

There is also evidence to link xenoestrogens to a range of human medical concerns, particularly reproductive problems such as reduced sperm count in men and breast cancer in women.

Even the "safest" herbicides such as Roundup (glycophosphate) – the second most widely used in the USA - are now known to pose a danger to wetland ecologies, and can totally decimate frog populations at routine contamination levels.

10.9 SCOPE OF ORGANIC FARMING

The Organic Farming Association of India (OFAI) is the country's only organization of grassroots organic farmers. Since Indian agriculture continues to remain a source of livelihood for mostly small farmers and peasants, OFAI membership reflects this ground reality as well.

OFAI is also committed to active involvement of women farmers in the decision-making structures of the association. Such involvement is mandatory and reflected in the organisation's bye-laws.

The association — which is registered under the Indian Societies Registration Act — was formed three years ago. Its memorandum of association was written and approved after a wide consultation with organic farmers. Unlike other organic farm certification systems, OFAI farm certification is done through the agency of trained organic farmers themselves. OFAI does not accept farm inspectors who do not themselves practice organic agriculture.

As OFAI farming is based on natural principles, it is firmly opposed to the introduction of Genetically Modified Organisms (GMOs) in agriculture and will actively campaign against such agriculture.

The ultimate objective of the association is to produce poison-free food for Indian consumers and to achieve this by maintaining the living fertility of Indian soils.

10.10 ORGANIC FARMING IN INDIA

Organic farming was practiced in India since thousands of years. The great Indian civilization thrived on organic farming and India was one of the most prosperous countries in the world, till the British ruled it.

In traditional India, entire agriculture was practiced using organic techniques, where the fertilizers, pesticides, etc., were obtained from plant and animal products. Organic farming was the backbone of the Indian economy and cow was worshipped (and is still done so) as God. The cow, not only provided milk, but also provided bullocks (for farming) and dung (which was used as fertilizers).

Organic farming is either really expensive or really cheap, depending on where you live and whether or not you are certified. Not only are the "natural" pesticides and fertilizers increasingly marketed by agribusiness as costly or costlier than their chemical counterparts, but proving you are an organic farmer requires certification, which is time-consuming and expensive. In the USA, converting to organic agriculture is a huge undertaking for commercial farmers, who have relied on chemical fertilizers and pesticides for many decades, but in India, the conversion is no less arduous, and far more ironic.

India's farmers are still mostly practicing organic methods, passed down for millenia. Organic fertilizer and natural pest control are the only tools available to most of these farmers, who have always lacked the financial resources to explore chemical solutions. But these farmers, whose produce is as organic as they come, cannot afford to pay the fees required to gain official certification.

As the international community adopts standards for organic agriculture, the challenges faced by farmers in the USA versus farmers in India in order to adapt are very different indeed. The danger is that the well-intentioned global move towards organic standards will make small organic farmers in countries like India, who have been never done anything but organic farming, no longer able to sell their crops.

10.10.1 In Rsponse to the $26 Billion Global Market for Organic Foods

The Indian Central Government set up a National Institute of Organic Farming in October 2003 in Ghaziabad, Madhya Pradesh. The purpose of this institute is to formulate rules, regulations and certification of organic farm products in conformity with international standards. The major organic products sold in the global markets include dried fruits and nuts, cocoa, spices, herbs, oil crops, and derived products. Non-food items include cotton, cut flowers, livestock and potted plants.

J.S. Mann, commissioner of Horticulture for the Union Agriculture Ministry, said, "The institute, set up as part of the national program for organic production, will have its offices across the country and is appointing certifying agencies for organic farm products for the domestic market."

The certifying agencies thus far named by the Centre include the APEDA (Agricultural and Processed Food Products Export Development Authority), the Tea Board, the Spices Board, the Coconut Development Board and the Directorate of Cashew and Cocoa. They will be accountable for confirming that any product sold with the new "India Organic" logo is in accordance with international criteria, and will launch major awareness and marketing campaigns, in India and abroad.

Rajnath Singh, Additional Director-General of the Indian Council of Agriculture Research (ICAR), in the LBS seminar on Organic Farming, said that currently the export of organic products is allowed only if "the produce is packed under a valid organic certification issued by a certifying agency accredited by a designated agency."

10.10.2 Board

Organic farming has been identified as a major thrust area of the 10th plan of the central government. 1 billion rupees have been allocated to the aforementioned National Institute of Organic Farming alone for the 10th five-year plan, Mann said. And by the end of 2004, according to APEDA chairman K.S. Money, 15% of farm products will be organically grown & processed. A working group has been set up by the Planning Commission, and the Department of Commerce has established National Organic Standards.

10.10.3 Most of India's Farms are "Organic by Default."

The irony and difficulty of the new governmental push for organic agriculture is that 65% of the country's cropped area is "organic by default," according to a study by Rabo India. By this somewhat degrading term they mean that small farmers, located mostly in the Eastern and North-Eastern regions of the country, have no choice except to farm without chemical fertilizers or pesticides. Though this is true in many cases, it is also true that a significant number of them have chosen to farm organically, as their forefathers have done for thousands of years.

Many have seen for themselves the effects of chemical farming – soil erosion and loss of soil nutrients, loss of nutrition in food, and human diseases resulting from the chemicals that inevitably seep into the water table, all the reasons for the urgent demand for organic foods and farming.

In 2002, according to Government statistics, from a total food production of over 200 million tonnes, the country produced only 14,000 tonnes of organic food products. India currently has only 1,426 certified organic farms.

This statistical discrepancy reveals that the weak link in the organic/economic chain is certification. Under current government policy, it takes four years for a farm to be certified as organic. The cost of preparing the report is a flat fee of Rs. 5000, and the certificate itself costs another Rs. 5000. While these costs are bearable for the new industrial organic greenhouses, they are equal to or more than an entire year's income for the average small farmer, if the costs of travel and inspection are included.

10.10.4 U.S. Dept. of Agriculture

In the United States, an organic farm plan or organic handling plan must be submitted to a USDA – accredited private or state certification program. The plan must explain all current growing and handling methods, and any materials that will be used – in the present, and any future plans must be included as well. Records for the last five years must be presented. Land must be chemical-free for three years prior to harvest, so a conventional farmer cannot receive the organic label for the transitional years. This will generally mean a decrease in income– crops may be less plentiful than with conventional fertilizers and pesticides, and yet the higher price for organic products won't yet be possible. Many farmers cannot afford the transition, even if they want to.

10.11 ORGANIC AGRICULTURAL MOVEMENTS

One solution to the small farmer's dilemma of how to both certify and survive is that of community certification. At the World Organic Congress, hosted last year by IFOAM (International Forum for Organic Agricultural Movements) in Victoria, Canada, the theme was "Cultivating Communities." The idea of community certification of organic farms was the main topic of discussion, a concept increasingly popular among farming communities worldwide who have become fed up with accreditation agencies.

In community certification, communities, on a non-profit basis, take charge of the certification process themselves. They evaluate the farmer's commitment to the stewardship of the soil, and examine from many angles whether the food is being grown in an environmentally sensitive way or not, rather than technical standards.

10.11.1 Cashew & Cocoa

While community certification may be a viable solution on the local level, it is our opinion that, in the global marketplace, less than exact technical standards will never be enough for today's consumer – and, in today's largely poisoned environment, it shouldn't be, either. Furthermore, such "soft" guidelines can easily backfire on the farmers themselves, as a system not based on facts must be by definition subject to local politics, bribery, favoritism, etc.

Certification to International Organic Standards will not be Easy for India's Small Farmers

India must find a way to keep the strict international organic standards intact if it wants to compete in the international market for organic foods– but is there a way to do it without leaving small farmers out in the cold? One obvious solution is for the government so eager to make India organic to subsidize these certification fees enough to make it a viable option for ordinary farmers, not just for neo-organic factory farms and greenhouses. Banks also could provide a more level playing field for small farmers– currently, almost all bank loans are for pure crop farmers, that is, monoculturalists. While many of these big-business farmers use harmful chemicals and processes, small farmers fertilizing their soil with recycled organic wastes are usually ineligible for insurance, much less state subsidies.

In the Hindu newspaper's annual environmental report, P.V. Satheesh, Director of the Deccan Development Society, writes, "It's a sobering thought that the farmers producing the best and cleanest food must pay extra to certify, instead of inorganic foods being certified as potentially bad for our health."

10.11.2 Shift to Chemical Farming in 1960s

During 1950s and 1960s, the ever increasing population of India and several natural calamities lead to severe food scarcity in the country. As a result, the government was forced to import food grains from foreign countries. To increase food security, the government had to drastically increase food production in India. The Green Revolution (under the leadership of M. S. Swaminathan) became the government's most important program in the 1960s. Several hectares of land was brought under cultivation. Hybrid seeds were introduced. Natural and organic fertilizers were replaced by chemical fertilizers and locally made pesticides were replaced by chemical pesticides. Large chemical factories such as the Rashtriya Chemical Fertilizers were established.

Before the Green Revolution, it was feared that millions of poor Indians would die of hunger in the mid 1970s. However, the Green Revolution, within a few years, showed its impact. The country, which greatly relied on imports for its food supply, reduced its imports every passing year. In 1990s, India had surplus food grains and once again became an exporter of food grains.

As time went by, extensive dependence on chemical farming has shown its darker side. The land is losing its fertility and is demanding larger quantities of fertilizers to be used. Pests are becoming immune, requiring the farmers to use stronger and costlier pesticides. Due to increased cost of farming, farmers are falling into the trap of money lenders, who are exploiting them no end, and forcing many to commit suicide.

Both consumers and farmers are now gradually shifting back to organic farming in India. It is believed by many that organic farming is healthier. Though the health benefits of organic food are yet to be proved, consumers are willing to pay higher premium for the same. Many farmers in India are shifting to organic farming due to the domestic and international demand for organic food. Further stringent standards for non-organic food in European and US markets have led to rejection of many Indian food consignments in the past. Organic farming, therefore, provides a better alternative to chemical farming.

According to the International Fund for Agriculture and Development (IFAD), about 2.5 million hectares of land was under organic farming in India in 2004. Further, there are over 15,000 certified organic farms in India. India, therefore is one of the most important suppliers of organic food to the developed nations. No doubt, the organic movement has again started in India.

10.11.3 The Official Position

As per a Food and Agriculture Organisation (FAO) study of mid-2003, India had 1,426 certified organic farms producing approximately 14,000 tons of organic food / produce annually. In 2005, as per Govt. of India figures, approximately 190,000 acres (77,000 hectares) were under organic cultivation. The total production of organic food in India as per the same reference was 120,000 tons annually, though this largely included certified forest collections.

10.11.4 Another Side to the Story

There are a number of farms in India which have either never been chemically-managed / cultivated or have converted back to organic farming because of their farmers' beliefs or purely for reason of economics. These thousands of farmers cultivating hundreds of thousands of acres of land are not classified as organic though they are. Their produce either sells in the open market along with conventionally grown produce at the same price or sells purely on goodwill and trust as organic through select outlets and regular specialist bazaars. These farmers will never opt for certification because of the costs involved as well as the extensive documentation that is required by certifiers.

10.11.5 New Potential Areas

About 65% of India's cropped area is not irrigated and it can be safely assumed that high-input demanding crops are not grown on these lands. Fertiliser use on drylands is always less anyway as chemical fertilisers require sufficient water to respond. Pesticide use in these lands would also be less as the economics of these hardy or "not-so profitable" crops will not permit expensive inputs. These areas are at least "relatively organic" or perhaps even "organic by default". While neither of these terms necessarily denotes a healthy farm or a recommended agriculture system, it would at least imply a non-chemical farm that can be converted very easily to an organic one providing excellent yields and without the necessity and effort of a lengthy conversion period.

10.12 ORGANIC FARMING

10.12.1 Economic Benefits

- Reduction in the use of external inputs and increase in output of organic produces with greater potential to benefit the health of farmers and consumers.

- More Productivity through the incorporation of natural process like natural cycles, nitrogen fixation and pest-predator relationship into the agricultural production process.

- Greater productive use of the biological and genetic potential of plant and animal species.

⊙ Long term sustainability of production levels.

⊙ I-Profitable and efficient production with emphasis on improved management and conservation of soil, water, energy and biological resources.

10.12.2 Ecological Benefits

⊙ Organic farming is much better for the environment as the energy consumption is much less than in the chemical farming.

⊙ It also uses less manurial inputs and completely avoids the synthetic fertilizers which otherwise pollute the soil, water and air.

⊙ It promotes biodiversity and a great variety of animals and plant interaction on earth.

⊙ Organic farmers focus on preserving the habitats of all species and their surrounding environment incuding the air and water.

⊙ Organic farming releases much less carbon dioxide than other farming system.

10.12.3 Social Benefits

⊙ Organic farming practices can be adopted in small farms and benefit marginal farmers.

⊙ It could reduce dependency on external inputs and costly technologies thus reducing the competitiveness and disparity among the farmers in a community.

⊙ It will also lead to food security at the family level and national level.

⊙ Organic farming is revival of a culture and brings back the indigenous knowledge, beliefs and value system that are almost on extinction now. It also contributes to employment generation at the community level.

10.13 ORGANIC AGRICULTURE – THE EXPERIENCE OF INDIA

India had developed a vast and rich traditional agricultural knowledge since ancient times and presently finding solutions to problems created by over use of agrochemicals. Present days' modern farming is not sustainable in consonance with economics, ecology, equity, energy and socio-cultural dimensions. Indiscriminate use of chemical fertilizers, weedicides and pesticides has resulted in various environmental and health hazards along with socio-economic problems.

Chemical base farming system is no more beneficial as it requires high input and low return, resulting migration of youth from rural area to urban area in search of other jobs. Besides that cultivable area and forest land is shrinking day by day and become biggest threat to habitat of animals and birds. Though agricultural production has continued to increase, but productivity rate per unit area has started to decline. The entire agricultural community is trying to find out an alternative sustainable farming system, which is ecologically sound, economically and socially acceptable. Sustainable agriculture is unifying concept, which considers ecological, environmental, philosophical, ethical and social impacts, balanced with cost effectiveness.

The answer to the problem probably lies in returning to our own roots. Traditional agricultural practices, which are, based on natural and organic methods of farming offer several effective, feasible and cost effective solutions to most of the basic problems being faced in conventional farming system. There is also need to conserve our traditional seed, some of which have drought resistant properties and resistant to different pest and diseases.

Looking these demerits of present day farming work was carried out on ancient organic farming practices viz; Biodynamic, Panchgavya, Rishi Krishi and Homa farming following objectives and brief of the achievement is given as under:

1. Quantum production equal or higher what is expected from optimum combination of agrochemicals.

2. Input generation at the farm

3. Continuous improvement in physico-chemical and biological properties of soil.

4. Par excellence produce quality with respect to nutrition, essential constituents, therapeutic value and storability.

5. Eco friendly and cost effective technology.

10.13.1 Achievements

Analysis of initial soil samples revealed that the organic carbon, available P and K and population of mould and bacteria were 0.53 per cent, 8.66 and 140 ppm and 1.3x104 and 3.7x106 cfu g-1, respectively.

After three years of organic cultivation, 3-4 fold improvement in physical, chemical and biological properties of the soil was. Compost prepared at the farm almost 2 to 3 times more than normal compost. Liquid manures/pesticides were prepared from leaves of leguminous tree and neem, caster leaves and other medicinal plant parts were found to have insecticidal and fungicidal properties.

Organic preparations contained high population of beneficial microbes.

Fruits

In a long term experiment, improvement in mango, guava, Indian gooseberry and papaya yield and quality was recorded with improvement in soil health.

Vegetables

In another experiment cauliflower, cabbage, okra and cowpea were grown in different organic package of practice. Yield, quality and cast benefit ratio was recorded maximum in comparison to application of recommended dose agrochemicals. Note: Input generated at the farm without any purchase but its cost was included while computing CB ratio at market rate.

Crop Protection

Neem and Pongamia based biodynamic pesticides found effective management of pests.

Diseases

Use of biodynamic pesticides and other organic preparations were also found effective in control of soil born diseases and brought immunity against adverse condition viz; frost and drought.

Physiological Disorder

Clustering (seedless fruit setting due low population of pollinators) disorder in mango causes 60 to 80 per cent yield in conventionally managed orchards but it was not noticed at the farm which was organically managed for a long time.

Validation of Technology Farmer's Fields

Thirty seven selected farmers in one area adopted the technology for cultivation of medicinal crops and earned profit from Rs.3, 45, 00.00 to Rs.4, 36,000.00 ha-1 and in other area farmers cultivated vegetables crops and earned profit from Rs.20, 000.00 to Rs.36,000.00 ha-1 over conventional cultivation.

Impact

- ◉ Farmers are engaged in production of inputs at their farm. Therefore cost of cultivation is reduced as compared to spraying of pesticides/fungicides to control pest and diseases.

- ◉ Looking the profit per unit area other farmers who are not trained in the programme are willing to join the practice.

- ◉ Because of better taste and flavour, produce are sold at premium price in the markets.

- ◉ At Azamagarh, all the produce are being purchased by organic India at higher rates for export because fields are organically certified.

10.13.2 Organic Production of Seeds and Planting Materials

Seed, seedlings and plantings materials are crucial, critical and essential input in agriculture production system. In conventional farming system the diseases are generally controlled by seed treatment, but it is not an option in organic agriculture. Keeping these views in mind, a new method for organic propagation by use of poly and net houses to improve the efficiency and extension in propagation period in a year has been standardized.

Biodynamic liquid manure/pesticides were used for the control of pests and plant vigour promotion and organic multiplication of horticultural planting materials has been standardized. Comparison of pollinator's activities at organic, conventional and farmer's conventional fields Comparative assessment of pollinators was made on pollinator populations in mango in different environments viz., organic farm, conventional orchard and farmers' fields. Observation on insect pollinator populations at different locations and environments revealed that the pollinator populations were adequate at organic farm.

10.13.3 Revenue Generation

Revenue generation was recorded more at organic farm compared same farm which was maintained conventionally earlier.

10.13.4 Other Activities

- Package of practiced developed for organic cultivation of horticultural crops
- Consultancy to government and non Government organization.
- Technical expertise and human resource development activity to sate department states of north-eastern region of India.
- Two hundred and fifty farmers are engaged in biodynamic production of medicinal crops such as basil, ashwagandha etc. on certified field.
- Farmers and state government have started biodynamic farming practices on their farms and gardens inconsultation.
- large number of progressive farmers have started organic cultivation of horticult ral crops after exposure visit to organic farm.
- Sale of organic mango, guava and seeds / planting materials to the farmers
- Training/ demonstrations activities are in progress since 2000.
- Seven states in NEH region of India have been identified to promote organic farming under Technology Mission Project of Govt. of India. On and off campus training and advices are being provided to Horticultural/ Agricultural Officers and farmers.

10.14 ORGANIC FARMING STEPS TO A SUCCESSFUL ORGANIC TRANSITION

The transition from conventional to organic farming requires numerous changes. One of the biggest changes is in the mindset of the farmer. Conventional approaches often involve the use of quick-fix remedies that, unfortunately, rarely address the cause of the problem. Transitioning farmers generally spend too much time worrying about replacing synthetic input with allowable organic product instead of considering management practices based on preventative strategies. Here are a few steps new entrants should follow when making the transition to organic farming:

(A) **Understand the Basics of Organic Agriculture and the Organic Farming Standards**—Since organic production systems are knowledge based, new entrants and transitional producers must become familiar with sound and sustainable agricultural practices. Transitional producers should be prepared to read appropriate information, conduct their own trials and participate in formal and informal training events. As mentioned, switching from conventional to organic farming is more than substituting synthetic materials to organic allowed materials. Organic farming is a holistic system that relies on sound practices focused on preventative strategies. Since there are often few organic remedies available to organic producers for certain problems, prevention is the key element in organic production.

(B) **Identify Resources that will Help You**—Existing organic farmers are generally very helpful in sharing valuable technical information. A good mentor should be able to provide transitional producers with knowledge, practical experience and suggest appropriate reading materials. Mentors are able to identify some of the most important challenges transitional farmers will be confronted with. Mentors may also help source production materials that are otherwise difficult to find. Producers should also contact agrologists, veterinarians and other agricultural and financial consultants, in order to learn ways to improve their current farming practices.

The Internet is a valuable source of information, especially to new organic farmers. A broad range of reading materials are available from many organic/ ecological organizations such as the Organic Agriculture Centre of Canada (OACC), the Atlantic Canadian Organic Regional Network (ACORN), the Canadian Organic Growers (COG), the Certified Organic Associations of British Columbia (COABC), the National Sustainable Agriculture Information Services/Appropriate Technology Transfer for Rural Areas (ATTRA), the Sustainable Agriculture Research and Education (SARE), and the Agri-réseau/agriculture biologique- Quebec. Consider joining an organic organization or network to access these valuable resources and establish good working contacts.

(C) **Plan your Transition Carefully**—Develop a transitional plan with clear and realistic goals. The plan should clearly identify various steps to be taken in making the transition to organic and be sure to include realistic timeframes. Identify your strengths and weaknesses. Consider ways to address any weaknesses, while building on strengths. The business side of the transitional plan should contain a multiple year budget and an effective/realistic marketing strategy. Make sure your list of expenses is comprehensive. Include all prerequisites to begin the transition; such as, mechanical weeding equipment, specialized composting equipment and applicators, additional handling equipment dedicated to the organic products, and processing equipment. Although the demand for organic products is continually growing, growers need to make sure they have a reliable market for the organic products they plan to produce.

Careful planning is very important. During the early part of the transitional period, yields are often depressed and premium prices for certified organic products are generally not yet obtainable. Use realistic yields and prices when evaluating the feasibility of your project.

In some instances, it is preferable to continue using conventional measures early on in the transitional process in order to avoid dramatic yield reduction which could jeopardize the financial well-being of the operation. Farmers who are planning to convert their livestock operation should consider certifying their fields first. This allows time to learn more about organic livestock management requirements while, at the same time, starting to produce organic feeds.

Although organic certifiers generally want to see the entire farm become organic, certifiers generally allow new entrants several years of transition time before the whole farm is fully certified.

Parallel production is the simultaneous production, processing or handling of organic and nonorganic crops, livestock and other products of a similar nature. Although this type of activity is highly discouraged by certifiers, some allow it, especially during the transition period. If permitted to practice parallel production, producers must be prepared to deal with significant record keeping in order to ensure traceability and organic integrity.

(D) **Understand your Soils and Ways to Improve Them**—Since soil is the heart of the organic farming system, it is crucial that new entrants understand the various characteristics and limitations of the soils found on their farm. Soil suitability may vary significantly from one field to the next. Fields with good drainage, good level of fertility and organic matter, adequate pH, biological health, high legume content, and with less weed and pest pressure, are excellent assets. Often these fields are the first ones ready for transition and certification.

Many tools exist to assess soils. Soil chemical, physical and biological analyses, soil survey and legume composition field assessments, and field yield histories are very important and should be considered early in the transition. Unhealthy soils require particular attention.

If farmers plan to grow crops without raising any livestock, it may be necessary for them to source allowable soil amendments such as composted manure, limestone, rock dust, and supplementary sources of nitrogen, phosphorus, potassium and micro-nutrients. Even with the best of crop rotations that include green manure crops like legumes (nitrogen fixing crops), transitional growers will be challenged if they want to obtain optimal yields without additional livestock manure, compost and/or other off-farm soil inputs. When these inputs are scarce or expensive, producers may benefit from integrating livestock on their farm.

Let's not forget, under organic production, farmers must be able to recycle nutrients through proper nutrient management practices: recycling through good manure and compost utilization, crop rotations, cover crops (green manure, catch, and nitrogen fixing crops), and by reducing nutrient losses due to leaching, over-fertilization, as well as poor manure and compost management (storage, handling, and spreading).

(E) **Identify the Crops or Livestock Suited for Your Situation**—Before growing a crop or raising any livestock, consider the following: degree of difficulty to grow or raise the product organically, land and soil suitability, climate suitability, level of demand for the product, marketing challenges, capital required, current prices for conventional, transitional and organic products, and profitability over additional workload.

(F) **Design Good Crop Rotations**—Once the crops are chosen, carefully plan the crop rotation(s) and select the most suitable cover crops (green manure, winter cover crops, catch crops, smother crops, etc.). Crop rotations are extremely important management tools in organic farming. They can interrupt pest life cycles, suppress weeds, provide and recycle fertility, and improve soil structure and tilth. Some rotational crops may also be cash crops, generating supplemental income.

On some farms, land base availability may be a limiting factor when planning your crop rotations. The transitional plan should, therefore, include crop rotation strategies. Responding to external forces such as new market opportunities may also have a significant impact on crop rotations, so farmers need to consider the effect that growing new crops has on their crop rotations and land base availability.

(G) **Identify Pest Challenges and Methods of Control**—It is important to know the crop's most common pests, their life cycles and adequate control measures. For instance, Colorado potato beetle may be a pest of significant importance when growing potatoes; cucumber beetles in cucurbitaceous crops (cucumber, squash, and melons); flea beetle in many seedlings crops; clipper weevil and Tarnish Plant Bug in strawberry crops.

There are several measures available to reduce pest pressure: crop rotation, variety selection, sanitation, floating row covers, catch crops, flamers, introduction of beneficial insects, bio pesticides, and inorganic pesticides. Transitional growers should be prepared to use and experiment with some of these options. When considering a new type of production, discuss pest issues with your agrologists, IPM specialists and/or other existing organic producers to optimize your chances of success.

Availability of organic supplies has improved significantly over the past few years. New pest control products containing B.t., spinosad, kaolin clay are effective and currently available to organic growers. It is often reported that the types of weeds found on the farm evolve with time as growers change the way they grow their crops and control their weeds. By keeping track of the weed population, growers will be able to refine their crop rotations and improve their control measures.

Under organic livestock management, cattlemen must provide attentive care that promotes health and meets the behavioral needs of various types of livestock. With good herd health practices, farmers rarely need to rely on conventional medicine. Organic cattlemen should, however, try to familiarize themselves with alternative remedies such as herbal/aroma therapies, homeopathy,and immune system promoters.

(H) **Be Ready to Conduct Your Own On-Farm Trials**—Successful organic farmers continuously try new and/or innovative management practices. Practices such as cover cropping, inter-planting, and use of various soil and pest control materials need to be evaluated regularly by organic farmers. Be prepared to try new approaches.

(I) **Be Ready to Keep Good Records**—Record keeping is one of the most important requirements to maintain organic integrity. Farmers are expected to keep detailed production, processing and marketing information. This information includes everything that enters and exits the farm. Third party, independent inspectors require farmers to present the above mentioned documentation when inspecting the farm operation. Once the record-keeping requirements are understood and the reporting procedure established, paperwork becomes routine.

(J) **Avoid These Common Mistakes**

◉ Underestimating the need for good transitional and marketing plans.

◉ Underestimating the need to fully understand the Organic Standard. Organic producers must understand the standard in order to know what is permitted and prohibited.

◉ Failing to think prevention. Transitional farmers should consider improving their crop rotation, soil and crop management skills, livestock management practices (feeding program, heard health program, grazing system, housing facilities, and husbandry).

10.15 ORGANIC FARMING INDIA'S FUTURE PERFECT

A budding interest in organic food offers farmers soaring incomes and higher yields, but critics say it's not the answer to India's fast-rising food demands

An Indian farm labourer displays a cabbage grown on an organic farm in India's Gujarat state.

India's struggling farmers are starting to profit from a budding interest in organic living. Not only are the incomes of organic farmers soaring – by 30% to 200%, according to organic experts – but their yields are rising as the pesticide-poisoned land is repaired through natural farming methods.

Organic farming only took off in the country about seven years ago. Farmers are turning back to traditional farming methods for a number of reasons.

First, there's a 10% to 20% premium to be earned by selling organic products abroad and in India's increasingly affluent cities, a move towards healthy living and growing concern over toxic foods and adulteration plaguing the food market.

Second, the cost of pesticides and fertilisers has shot up and the loans farmers need to buy expensive, modified seed varieties are pushing many into a spiral of debt. Crippling debt and the burden of loans are trriggering farmer suicides across the country, particularly in the Vidarabha region of Maharashtra. Organic farming slashes cultivation and input costs by up to 70% due to the use of cheaper, natural products like manure instead of chemicals and fertilisers.

Third, farmers are suffering from the damaging effects of India's green revolution, which ushered in the rampant use of pesticides and fertilisers from the 1960s to ensure bumper yields and curb famine and food shortages. Over the decades, the chemicals have taken a toll on the land and yields are plunging.

"Western, modern farming has spoiled agriculture in the country. An overuse of chemicals has made land acidic and hard, which means it needs even more water to produce, which is costly," "Chemicals have killed the biggest civilisation in agriculture – earthworms, which produce the best soil for growth."

Umesh Vishwanath Chaudhari, 35, a farmer in the Jalgaon district in Maharashtra, switched to organic farming seven years ago after experiencing diminishing yields from his 8-hectare (20-acre) plot. He came across a book on organic farming techniques using ancient Vedic science. He started making natural fertilisers and pesticides using ingredients such as cow manure, cow urine, honey and through vermicomposting – the process of using earthworms to generate compost. Since then, his yields and income have risen by 40%, and worms have returned to his soil. He sells lime, custard apple and drumsticks to organic stores in Pune, Mumbai and other cities, while his cotton is bought by Morarka, a rural NGO.

He plans to convert another 2 hectares to organic cotton and buy 10 cows to make his own manure, rather than buying it. "Using manure instead of pesticides and fertilisers has cut my costs by half, and I get a premium on these goods," he says. "I used to drive a scooter, but in the past few years I've been able to afford a bike and car – and even two tractors."

Udday Dattatraya Patil, 43, an agriculture graduate, turned to organic farming after his crops were showing a deficiency in feed, leading to rising fertiliser costs. In addition, his banana crop was being wrecked by temperature fluctuations and climate change. "Because bananas are sensitive to temperature change, 20% went to waste. Organic bananas can withstand this. Now none are wasted," he says. Now he has 40 cows and bulls whose manure he can use for fertiliser, as well as vermicompost units. His yields have increased by 20% and income by 30%.

Although he is hailed as a progressive agriculturalist by his fellow villagers, he is the only organic farmer out 3,000 in Chahardi, in Jalgaon district. "Some have tried but they give up if there aren't immediate results. Organic farming requires effort, and you have to invest in organic inputs," he adds.

Many farmers are reluctant to make the leap because they fear a drop in yields in the initial period; good results tend to show after three years. Moreover, the market is growing by 500% to 1,000% a year, according to Morarka, but it only represents 0.1% of the food market.

Kavita Mukhi organises a weekly organic farmers' market in Mumbai, where producers sell direct to consumers. She is trying to boost awareness about organic food. "The only way you hear about it is if you stumble on an organic shop," she says. "There's no widespread marketing or awareness of the benefits."

Once the awareness increases, organic agriculturalists believe more farmers will join the movement because it's favourable to small farmers. They already have the cows and buffalos needed to recycle biomass at the farm level, which is, essentially, the foundation of organic farming.

"Unlike Europe, India's modern farming revolution is not very old, meaning they still possess the knowhow for cultivation without modern chemical inputs," says Mukesh Gupta of Morarka.

While critics argue that organic farming is not the answer to India's rising food demands, those in favour say it's the only sustainable way out for impoverished farmers.

10.15.1 Organic Farming in India: Myths and Realities

Whether organic farming can address the multitude of problems faced by Indian agriculture at present is a major issue.

Whether organic farming can address the multitude of problems faced by Indian agriculture at present is a major issue

Agriculture in India is one of the most important sectors of its economy. It provides livelihood to almost two thirds of the work force in the country and accounts for 18% of India's GDP. About 43 % of India's geographical area is used for agricultural activity.

Agriculture is the single largest employment provider and plays a vital role in the overall socio-economic development of India. A large number of production systems are in practice in different parts of the country. Large scale use of inputs both organic and inorganic has been a common sight in many of the farming situations in the past several decades. However in recent times the concept of organic farming is being forcefully projected as the method for sustaining the agricultural production in the country.

Organic farming is a form of agriculture which avoids or largely excludes the use of synthetic fertilizers and pesticides, plant growth regulators, and livestock feed additives. Organic farming relies on crop rotation, crop residues, animal manures, bio-fertilizers and mechanical cultivation to maintain soil productivity, to supply plant nutrients, and to control weeds, insects, diseases and other pests.

Before jumping into organic farming bandwagon, we need to have answers to the following: What level of crop yield/ productivity is acceptable? Is it suitable for country like India with a large population to feed? Whether available organic sources of plant nutrients sufficient for pure organic farming? And, are organic farming technologies sustainable in long run?

Whether organic farming can address the multitude of problems faced by Indian agriculture at present is a major issue. Further, the virtues attributed to organic farming need to be rechecked before coming to any conclusions.

10.15.2 Issues of Concern

Organic farming and nutrient supply—At present, there is a gap of nearly 10 million tonnes between annual addition and removal of nutrients by crops which are met by mining nutrients from soil. A negative balance of about 8 million tonnes of NPK is foreseen in 2020, even if we continue to use chemical fertilizers, maintaining present growth rates of production and consumption. The most optimistic estimates at present, show that only about 25-30 per cent nutrient needs of Indian agriculture can be met by utilizing various organic sources.: These organic sources are agriculture wastes, animal manure etc.

Organic farming and plant protection—Plant protection against the ravages of pests, diseases and weeds is an important issue in any modern high production system. The exclusion of pesticides for plant protection poses greater risk of yield losses. The options available under organic production systems are very few and crop specific. Often they are very slow and the success rate depends on the prevailing weather conditions leading to low to moderate effectiveness even in the recommended crops and situations. Thus they limit the realization of full potential of crop yields. Any sudden outbreak of insect pests or plant disease can completely destroy the crops, unless requisite chemical pesticides are used.

Organic farming and crop productivity—In general, it is observed that the crop productivity declines under organic farming. The extent of decline depends on the crop type, farming systems practices followed at present etc. The decline is more in high yielding and high nutrient drawing cereals as compared to legumes and vegetables and in irrigated systems as compared to rainfed and dryland farming systems. Without using fertilisers, the requirement of area to merely sustain the present level of food grain production will be more than the geographical area of India! This is simply neither possible nor sustainable.

Organic farming and certification processes—Hitherto there are no standard certification processes uniformly applicable across different agro-climatic conditions. Both process and product certification procedures are still in evolutionary stage and need further progress before they can be effectively adopted. Due to biological nature of both processes and products, there is always an element of dynamism subject to temporal and spatial conditions. The presently available certification procedures are very cumbersome and expensive and out of reach for the common farmer. Given the highly fragmented holdings of the farmers, there is every possibility of "contamination" from the neighbouring farms – besides the temptation to use chemical inputs to boost yields.

Organic farming and heterogeneity of inputs—There is a large variability in the inputs used in organic farming. Due to biological nature of the inputs, prescribing uniform standards and maintaining them in different agro-climatic conditions is beyond ones control. Thus, there can not be a common input recommendation as in fertilizers or pesticides. This leads to arbitrariness on the part of organic farmers as far as input management is concerned.

Organic farming and food quality—It is often opined that the quality of the organically produced food is superior to that of conventionally produced food. However, there is no such conclusive proof to justify the nutritional superiority of the organically produced food, over conventionally produced food. If the conventionally produced foods are blamed to contain traces of chemical residues, the organically produced foods are equally to be blamed for their contamination with harmful bacteria and other organisms inimical to the health of the consumers.

Organic farming products and marketing—There are no diagnostic techniques available as of now to distinguish products from different farming systems. The perceived belief that organic products are good for health is fetching them premium prices. However, unscrupulous hawkers may sell anything and everything as organically produced to unsuspecting buyer at higher prices resulting in outright cheating.

Organic farming and switch over period—A transition period of 3-4 years is generally required to convert a conventional farm into an organic farm. In this period, the produce is not considered as organically produced. The reduced yields and lack of benefits of premium for the produces is a double blow for the farmers leading to financial losses which are substantial for the small to medium farmers

10.15.3 The Possible Options

With all the above concerns, organic farming is not feasible as an alternative to conventional farming under all circumstances in Indian context. The shortfall in inorganic nutrient supply, uneconomic returns to inorganic inputs under dryland and rainfed farming systems, inherent better response to organic farming in crops like vegetables, legumes and millets under traditional farming systems paves way for integration of conventional farming with organic farming. Such integration on sound scientific basis will be effective in addressing the problems of micronutrient deficiencies; recycling of crop residues, farm wastes, rural and urban wastes; besides effectively meeting growing food demands of rising populations. There will also be scope for practicing organic farming on case to case basis in traditional strongholds like hilly areas, rain fed and dry land farming system to cater to the demands of organic produces in urban areas who would pay premium prices for such commodities.

Organic farming should be considered for lesser endowed region of the country. It should be started with low volume high value crops like spices and medicinal aromatic crops. A holistic approach involving integrated nutrient management, integrated pest management, enhanced input use efficiency and adoption of region-specific promising cropping systems would be the best farming strategy for India

Bottom-line—Organic foods are a matter of choice of the individuals or enterprises. If somebody wants to go in for organic farming, primarily on commercial consideration / profits motive, to take advantage of the unusually higher prices of organic food, they are free to do so. Organic farming is essentially a marking tool, and cannot replace conventional farming for food security, quality and quantity of crop outputs. With a growing population and precarious food situation, India cannot afford to take risk with organic farming alone.

10.16 FOOD SECURITY AND SAFETY

Food security refers to the availability of food and one's access to it. A household is considered food-secure when its occupants do not live in hunger or fear of starvation. According to the World Resources Institute, global per capita food production has been increasing substantially for the past several decades. In 2006, MSNBC reported that globally, the number of people who are overweight has surpassed the number who are undernourished - the world had more than one billion people who were overweight, and an estimated 800 million who were undernourished. According to a 2004 article from the BBC, China, the world's most populous country, is suffering from an obesity epidemic. In India, the second-most populous country in the world, 30 million people have been added to the ranks of the hungry since the mid-1990s and 46% of children are underweight.

Worldwide around 852 million people are chronically hungry due to extreme poverty, while up to 2 billion people lackfood security intermittently due to varying degrees of poverty. Six million children die of hunger every year - 17,000 every day. As of late 2007, export restrictions and panic buying, US Dollar Depreciation, increased farming for use in biofuels,world oil prices at more than $100 a barrel, global population growth, climate change, loss of agriculturall and to residential and industrial development, and growing consumer demand in China and India are claimed to have pushed up the price of grain. However, the role of some of these factors is under debate. Some argue the role of biofuel has been overplayed as grain prices have come down to the levels of 2006. Nonetheless, food riots have recently taken place in many countries across the world.

It is becoming increasingly difficult to maintain food security in a world beset by a confluence of "peak" phenomena, namely peak oil, peak water, peak phosphorus, peak grainand peak fish. Approximately 3.3 billion people, more than half of the planet's population, live in urban areas as of November 2007. Any disruption to farm supplies may precipitate a uniquely urban food crisis in a relatively short time. The ongoing global credit crisis has affected farm credits, despite a boom in commodity prices. Food security is a complex topic, standing at the intersection of many disciplines.

A new peer-reviewed journal of Food Security: The Science, Sociology and Economics of Food Production and Access to Food began publishing in 2009. In developing countries, often 70% or more of the population lives in rural areas. In that context, agricultural development among smallholder farmers and landless people provides a livelihood for people allowing them the opportunity to stay in their communities. In many areas of the world, land ownership is not available, thus, people who want or need to farm to make a living have little incentive to improve the land.

In the US, there are approximately 2,000,000 farmers, less than 1% of the population. A direct relationship exists between food consumption levels and poverty. Families with the financial resources to escape extreme poverty rarely suffer from chronic hunger, while poor families not only suffer the most from chronic hunger, but are also the segment of the population most at risk during food shortages and famines.

Two commonly used definitions of food security come from the UN's Food and Agriculture Organization (FAO) and the United States Department of Agriculture (USDA):

- ⊙ Food security exists when all people, at all times, have physical, social and economic access to sufficient, safe and nutritious food to meet their dietary needs and food preferences for an active and healthy life.

- ⊙ Food security for a household means access by all members at all times to enough food for an active, healthy life. Food security includes at a minimum (1) the ready availability of nutritionally adequate and safe foods, and (2) an assured ability to acquire acceptable foods in socially acceptable ways (that is, without resorting to emergency food supplies, scavenging, stealing, or other coping strategies).

The stages of food insecurity range from food secure situations to full-scale famine. "Famine and hunger are both rooted in food insecurity. Food insecurity can be categorized as either chronic or transitory. Chronic food insecurity translates into a high degree of vulnerability to famine and hunger; ensuring food security presupposes elimination of that vulnerability. [Chronic] hunger is not famine. It is similar to undernourishment and is related to poverty, existing mainly in poor countries.

10.16.1 Stunting and Chronic Nutritional Deficiencies

Children and a nurse attendant at a Nigerian orphanage in the late 1960's with symptoms of low calorie and protein intake.

Many countries experience perpetual food shortages and distribution problems. These result in chronic and often widespread hunger amongst significant numbers of people. Human populations respond to chronic hunger and malnutrition by decreasing body size, known in medical terms as stunting or stunted growth. This process starts in utero if the mother is malnourished and continues through approximately the third year of life. It leads to higher infant and child mortality, but at rates far lower than during famines. Once stunting has occurred, improved nutritional intake later in life cannot reverse the damage. Stunting itself is viewed as a coping mechanism, designed to bring body size into alignment with the calories available during adulthood in the location where the child is born. Limiting body size as a way of adapting to low levels of energy (calories) adversely affects health in three ways:

- Premature failure of vital organs occurs during adulthood. For example, a 50-year-old individual might die of heart failure because his/her heart suffered structural defects during early development.

- Stunted individuals suffer a far higher rate of disease and illness than those who have not undergone stunting.

- Severe malnutrition in early childhood often leads to defects in cognitive development.

"The analysis... points to the misleading nature of the concept of subsistence as Malthus originally used it and as it is still widely used today. Subsistence in not located at the edge of a nutritional cliff, beyond which lies demographic disaster. Rather than one level of subsistence, there are numerous levels at which a population and a food supply can be in equilibrium in the sense that they can be indefinitely sustained. However, some levels will have smaller people and higher normal mortality than others."

10.17 ACHIEVING FOOD SECURITY

"The number of people without enough food to eat on a regular basis remains stubbornly high, at over 800 million, and is not falling significantly. Over 60% of the world's undernourished people live in Asia, and a quarter in Africa. The proportion of people who are hungry, however, is greater in Africa (33%) than Asia (16%). The latest FAO figures indicate that there are 22 countries, 16 of which are in Africa, in which the undernourishment prevalence rate is over 35%."

By way of comparison, in one of the largest food producing countries in the world, the United States, approximately one out of six people are "food insecure", including 17 million children, according to the U.S. Department of Agriculture. Food insecurity is measured in the United States by questions in the Census Bureau's Current Population Survey. The questions asked are about anxiety that the household budget is inadequate to buy enough food, inadequacy in the quantity or quality of food eaten by adults and children in the household, and instances of reduced food intake or consequences of reduced food intake for adults and for children. A National Academy of Sciences study commissioned by the USDA criticized this measurement and the relationship of "food security" to hunger, adding "it is not clear whether hunger is appropriately identified as the extreme end of the food security scale."

In its "The State of Food Insecurity in the World 2003", FAO states that:

'In general the countries that succeeded in reducing hunger were characterised by more rapideconomic growth and specifically more rapid growth in their agricultural sectors. They also exhibited slower population growth, lower levels of HIV and higher ranking in the Human Development Index'.

As such, according to FAO, addressing agriculture and population growth is vital to achieving food security. Other organisations and people (e.g. Peter Singer) have come to this same conclusion, and advocate improvements in agriculture and population control.

USAID proposes several key steps to increasing agricultural productivity which is in turn key to increasing rural income and reducing food insecurity. They include:

- Boosting agricultural science and technology. Current agricultural yields are insufficient to feed the growing populations. Eventually, the rising agricultural productivity drives economic growth.

- Securing property rights and access to finance.

- Enhancing human capital through education and improved health.

- Conflict prevention and resolution mechanisms and democracy and governance based on principles of accountability and transparency in public institutions and the rule of law are basic to reducing vulnerable members of society.

The UN Millennium Development Goals are one of the initiatives aimed at achieving food security in the world. In its list of goals, the first Millennium Development Goal states that the UN "is to eradicate extreme hunger and poverty", and that "agricultural productivity is likely to play a key role in this if it is to be reached on time".

"Of the eight Millennium Development Goals, eradicating extreme hunger and poverty depends on agriculture the most. (MDG 1 calls for halving hunger and poverty by 2015 in relation to 1990.)

Notably, the gathering of wild food plants appears to be an efficient alternative method of subsistence in tropical countries, which may play a role in poverty alleviation.

10.17.1 The Agriculture-Hunger-Poverty Nexus

Eradicating hunger and poverty requires an understanding of the ways in which these two injustices interconnect. Hunger, and the malnourishment that accompanies it, prevents poor people from escaping poverty because it diminishes their ability to learn, work, and care for themselves and their family members. Food insecurity exists when people are undernourished as a result of the physical unavailability of food, their lack of social or economic access to adequate food, and/or inadequate food utilization. Food-insecure people are those individuals whose food intake falls below their minimum calorie (energy) requirements, as well as those who exhibit physical symptoms caused by energy and nutrient deficiencies resulting from an inadequate or unbalanced diet or from the body's inability to use food effectively because of infection or disease. An alternative view would define the concept of food insecurity as referring only to the consequence of inadequate consumption of nutritious food, considering the physiological utilization of food by the body as being within the domain of nutrition and health. Malnourishment also leads to poor health hence individuals fail to provide for their families. If left unaddressed, hunger sets in motion an array of outcomes that perpetuate malnutrition, reduce the ability of adults to work and to give birth to healthy children, and erode children's ability to learn and lead productive, healthy, and happy lives. This truncation of human development undermines a country's potential for economic development–for generations to come.

There are strong, direct relationships between agricultural productivity, hunger, and poverty. Three-quarters of the world's poor live in rural areas and make their living from agriculture. Hunger and child malnutrition are greater in these areas than in urban areas. Moreover, the higher the proportion of the rural population that obtains its income solely from subsistence farming (without the benefit of pro-poor technologies and access to markets), the higher the incidence of malnutrition. Therefore, improvements in agricultural productivity aimed at small-scale farmers will benefit the rural poor first.

Increased agricultural productivity enables farmers to grow more food, which translates into better diets and, under market conditions that offer a level playing field, into higher farm incomes. With more money, farmers are more likely to diversify production and grow higher-value crops, benefiting not only themselves but the economy as a whole."

Researchers suggest forming an alliance between the emergency food program and CSA Farms, as currently food stamps cannot be used at farmer's markets and places in which food is less processed and grown locally.

Food safety is a scientific discipline describing handling, preparation, and storage of food in ways that prevent food borne illness. This includes a number of routines that should be followed to avoid potentially severe health hazards. Food can transmit isease rom person to person as well as serve as a growth medium for bacteria that can cause food poisoning. Debates on genetic food safety include such issues as impact of genetically modified food on health of further generations and genetic pollution of environment, which can destroy natural biological diversity. In developed countries there are intricate standards for food preparation, whereas in lesser developed countries the main issue is simply the availability of adequate safe water, which is usually a critical item. In theory food poisoning is 100% preventable.

10.17.2 FAO Trust Fund for Food Security and Food Safety

The World Food Summit, held in November 1996, was the first global gathering at the highest political level to focus solely on food security. In adopting the Rome Declaration on World Food Security and the World Food Summit Plan of Action, it renewed the commitment of the international community to ensuring food for all. The Declaration enunciates both the ultimate goal and the immediate target: "We pledge our political will and our common and national commitment to achieving food security for all and to an ongoing effort to eradicate hunger in all countries, with an immediate view to reducing the number of undernourished people to half their present level by 2015."

FAO has a major role to play in assisting countries in implementing the provisions of the World Food Summit Plan of Action that fall within its mandate, as well as in monitoring, through its Committee on World Food Security (CFS), overall progress in achieving the Summit's goals.

Against this background and in accordance with FAO's Financial Regulation 6.7, the Director-General established the FAO Trust Fund for Food Security and for Emergency Prevention of Transboundary Pests and Diseases of Animals and Plants with an initial target of US$500 million.

Areas of Intervention Covered by the FAO Trust Fund

- Food security
- Emergency prevention of transboundary animal and plant pests and diseases
- Assistance in project and programme studies to increase investment

This FAO Trust Fund will be an important source of demand-driven funding to supplement the present trust funds, which support key components of the Organization's Field Programme with emphasis on catalytic projects addressing long-term structural needs of the poor(70 percent of whom are in the rural sector) in the basic areas.

Several donors have contributed to the FAO Trust Fund for Food Security and Food Safety. The Italian Government has contributed 67 million euros out of 100 million euros committed. The funds have been earmarked for global food security, anti-disease and investment promotion projects.

Other donors include:

- ⊙ CO_2 Three SPFS projects in Cambodia, Haiti and Sierra Leone
- ⊙ **Greece:** Two projects in Armenia on Pesticide Quality Control and Abattoir Development and one to support Fisheries in the Eastern Mediterranean
- ⊙ **Czech Republic:** Three projects, in the Balkan Region, Mali and Morocco.

WHO

10.18 NUTRITION, FOOD SAFETY AND FOOD SECURITY

10.18.1 Food Security Monitoring System (FSMS)

FSMS gathers data every quarter from 665 households in rural areas of Tajikistan. For the second and fourth rounds of data collection (in January and July 2009), WHO/Europe worked closely with the World Food Programme (WFP) and WHO headquarters to incorporate nutrition indicators, including information on nutritional status and dietary diversity in children under 5 years and women aged 19–49. The regular collection of nutrition information in this sample (every 6–12 months) should provide information to improve the planning and targeting of programmes related to food security and nutrition.

Data are collected and managed by the nongovernmental organization CSR Zerkalo, the Ministry of Health, the Republican Centre for Nutrition and the Paediatric Surgery Centre in Tajikistan. Quarterly food security bulletins give the results from all data collection rounds.

10.18.2 Complementary Feeding Project in 6 Pilot Districts of Tajikistan

Iron deficiency anaemia and growth retardation are still major concerns among infants and young children in Tajikistan, especially in the highly deprived rural regions such as Khatlon and (Gorno-Badakhshan Autonomous Province) GBAO. In 2003 (Micro-Nutrient Status Survey), 37.6% of infants/children 6–59 months of age were found to be anaemic. Iron deficiency was present in 54% of the children with moderate and severe anaemia. According to the findings of the MICS 2005 the global chronic malnutrition rate in Tajikistan was 27% – with the highest rates in Khatlon and GBAO oblasts.

The WHO Global Strategy for Infant and Young Child Feeding recommends that infants should be exclusively breastfed up to 6 months of age. After the sixth month of life, breast-milk alone cannot guarantee the coverage of all nutrient requirements. Complementary feeding represents therefore a critical transition period of the early months and years of life during which infants and young children are particular vulnerable.

A review of the complementary feeding patterns in Tajikistan indicated that although most of the infants and children are breastfed, complementary feeding is not timely and adequate:

- The introduction of foods such as legumes, meat, liver, fruits and vegetables is later than advisable;

- Meat, milk and eggs etc are not adequately present in the diet;

- Black tea, which contains compounds that can interfere with iron absorption is commonly given during 6–24 months of age.

As a consequence, the diet of infants and children has poor diversity and is low in nutrient density.

Current approaches to the prevention of micronutrient deficiencies in infants include the provision of iron supplements and the use of fortified wheat flour. Supplementation is an expensive and human resource intensive action that does not seem cost-effective, due to poor compliance. Food fortification needs to be better targeted to this age group. Commercial preparations are available but hardly affordable. The long-term goal should be to optimize complementary feeding with available and affordable foods.

The pilot project was launched in January 2008 and finished in April 2009. The results of implementation were generated and included in the final report on October 2009.

The objective of the project in six districts of Khatlon and Gorno-Badakhshan was to implement an integrated food-based strategy for improving complementary feeding and nutritional status of infants and children 6–24 months of age and to evaluate the long term effectiveness and feasibility of an in-home fortification – using micronutrient powder (Sprinkles®) – to reduce anaemia in children.

The following activities are being carried out in Tajikistan as part of the project.

- Complementary feeding recommendations were drafted using linear programming approach to identify adequacy and costs of various combinations of locally available foods and propose the most cost-effective solution to fill nutrient gaps. The recommendations are currently being tested for feasibility and acceptability in the six project districts of Tajikistan.

- Effectiveness trial on the use of in-home fortification in Tajikistan. The use, acceptability and impact of in-home fortification through sprinkles are being tested in the six project districts of Tajikistan.

- Health workers continuously received training in health promotion, infant feeding, methods to assess nutritional status and dietary intake assessment.

10.19 INDIA'S FOOD SECURITY EMERGENCY

Corporate influence on food production and large, chemical monoculture farms is causing a severe food insecurity crisis.

The proposed introduction of the Food Security Act by the UPA Government is a welcome and much needed step towards securing the right to food for all of India's citizens. The right to food is the basis of the right to life, and Article 21 of the Constitution guarantees the right to life of all Indian citizens.

India has emerged as the capital of hunger, illustrated by the fact that per capita consumption has dropped from 178 kg in 1991 - the beginning of the period of economic reforms - to 155 kg in 200-2003.

Daily calorie consumption of the bottom 25 per cent of the population has decreased from 1683 k.cal in 1987-88 to 1624 k.cal in 2004-05, against a national norm of 2400 and 2011 k cal/day for rural and urban areas respectively.

Therefore, a response on the food security front is really a response to a national emergency. Unfortunately, the current approach to food security in the draft law Food Security Act ignores the larger food crisis.

10.19.1 Food Security

Food security includes three vital aspects: Ecological security, food sovereignty, and food safety.

Land, water and biodiversity are the natural capital for the food production. Currently, each of these is under severe threat. The land-grab of fertile farm land is not just an issue of injustice against farmers, but it is actually a threat to the nation's food security.

If fertile farm lands disappear, there will be no food.

India's seed wealth is being handed over to global corporations leading to erosion of biodiversity and undermining of farmers' rights. Without seed sovereignty there can no food sovereignty.

The country as a whole is growing increasingly vulnerable on the ecological security front, even more so because of climate change. That is why ecological agriculture that builds ecological security and resilience is necessary at this time for food security.

Food sovereignty is increasingly being lost as food and agriculture is hijacked by global agribusiness and determined by the unfair rules of WTO. That is why fair trade and WTO reform is vital to food security, and food sovereignty must be its foundation.

India's food is becoming unsafe and hazardous, with GMOs and chemically-processed food being promoted.

Corporations like Pepsi and Coke sit on the newly formed Food Safety Committee. The corporate influence on issues of safety is denying citizens their right to safe food. Without safe food there is no food security; without food democracy there is neither food safety nor food security.

The biggest blind spot in the dominant paradigm of food security is neglecting food production and food producers as a core element in the current food security approach.

You cannot provide food to people if you do not first ensure that food is produced in adequate quantities. And in order to ensure food production, the livelihood of food producers must be ensured. The right of food producers to produce food is the foundation of food security. This right has internationally evolved through the concept of "food sovereignty". In Navdanya we refer to it as Anna Swaraj.

Food sovereignty is derived from socio-economic human rights, which include the right to food and the right to produce food for rural communities.

10.19.2 Fresh Focus: Marketing Food Safety and Food Security

When making a presentation to a client or a potential client, they will ask about your food safety and security plans. Most companies are receiving visits by different independent third party auditing companies, so you might as well market what you are doing to address these important topics. When a decision-maker visits your facility to finalize a deal, these two points can make or break your business.

Take time today, before these visits, to imagine you are visiting your own facility for the first time. Drive in and take note of things. Were you stopped by a guard out front? Your food security plan should have someone posted to keep unwanted people out. Is anyone at your front desk when you come into your office area…or can you just wander in? Security includes your employees! Do visitors (and employees) need to wear identification badges? The checklist should go on and on.

10.20 STRIVE FOR NO SURPRISES

When hosting a visitor in your facility, they will observe small things that you just won't see on a day-to-day basis. The best way to avoid surprises is to be prepared everyday by having good procedures. Start before entering the production area. Everyone should wash their hands and don hairnets and beard nets (if necessary). Then walk through the footbath(s) and into the production area. Your facility should be as clean as possible at all times so there are no surprises. This means the floors, equipment, fluorescent lights and ceiling tiles, too.

What does all this basic information have to do with marketing? It's one thing to say you have a food safety program in place. It's another to have your customer actually observe it. Marketing includes taking visitors on tours. What better time to identify your food safety and security procedures by actually observing them in action.

Talk to your customers (and this includes industry suppliers and customers as well as your product's eventual end-users) about what you are doing to provide the best in food safety so they have a greater level of confidence in your products. Your web site can talk to both industry clients AND to your end-users.

10.20.1 Ready Pac – Ready for Anything

One personal experience I had included being stopped outside the Ready Pac Produce, Inc. facility in Irwindale, CA, years before food safety and security became important. Remember, back in the 1990's, this was not an issue…but it was at Ready Pac. I wasn't allowed to park inside the fenced-in area adjacent to the building (in a driving rainstorm, by the way). I had to park pretty far away, as there were other visitors ahead of me that day.

Dennis Gertmenian, the founder and chairman of the company, explained it to me. He said that by having a guard outside the parking area, and having him not only check my identification, but also call into the office to be certain I did have an appointment with the person I said I was seeing, they were eliminating risk. Dennis was always ahead of the curve. Once inside, I was escorted around the plant to see exactly what else they were doing to keep their produce safe.

Consumers assume the food they buy is safe. In nearly every case, it is. But why not take the extra steps necessary to ensure the public that you are doing everything possible to provide them with safe produce. Educate them! If you don't believe me, just check out the Ready Pac web site (http://www.readypac.com). Right on their home page it says "Food Safety Comes First", then below that it asks you to, "Click here for the latest update on our Food Safety Program."

10.20.2 Commitment to Quality – Not Just a Slogan Anymore

When we talk about quality, most people think quality of the produce. Food safety is a very important aspect of achieving quality. From the raw materials that are procured to the people on the cutting and packing lines, quality needs to be stressed from top management down. Dennis Gertmenian said it very well – "We will provide the highest quality fresh-cut produce on a national basis. Our valued associates are innovative and driven to exceed the expectations of our customers and consumers."

Advertisers are also touting food safety. Just take a look at some of the headlines from last month's Fresh Cut magazine. "Food Safety Starts While Crops Are Being Grown," touts one ad. "Automation of Water Disinfection. Essential for HACCP & Food Safety." said another. A third tells you that "Cleaner Cuts, Longer Life, Happier Customers" are important.

11

IMPACT OF THE COVID-19 PANDEMIC ON AGRICULTURE AND THE RURAL ECONOMY

11.1 INTRODUCTION: AN OVERVIEW

The novel Coronaviruspandemic has rapidly spread across the world, adversely affecting the lives and livelihoods of millions across the globe. India reported its first infection on 30 January 2020, prompting the authorities to soon initiate various measures to contain the spread of the epidemic. Given that the disease is highly contagious, the much-needed nation-wide lockdown was enforced starting 25 March 2020 in order to contain the spread of COVID-19 pandemic. During the initial few weeks, the restrictions were strict and all non-essential activities and businesses, including retail establishments, educational institutions, places of religious worship, across the country were prohibited from operating. Subsequently, these restrictions are being gradually eased in a phased manner in most parts of the country.

The agricultural & allied sector carries immense importance for the Indian economy. It contributes nearly one-sixth to the Indian national income and provides employment to nearly 50% of the workforce. It is fundamental for ensuring food security of the nation and also influences the growth of secondary and tertiary sector of the economy through its forward and backward linkages. The performance of agricultural sector greatly influences achievements on many other fronts. For instance, World Development Report 2008 released by World Bank emphasises that growth in agriculture is, on average, at least twice as effective in reducing poverty as growth outside agriculture. Agricultural growth reduces poverty directly, by raising farm incomes, and indirectly, through generating employment and reducing food prices. In other words, a thriving agricultural sector is a boon for most sectors of the Indian economy.

11.2 GLOBAL VS NATIONAL YIELD OF MAJOR CROPS

Although, India is one of the largest producer of some of the agriculture and horticulture products, yet the national yield of major crops (except ground nut) is less than the global average yield production. Further, the national yield of such crops is far less than the highest yield achieved in other parts of the world. (Table1). The COVID 19 pandemic has adversely impacted the globally agriculture sector and Indian agriculture sector is no exception.

Table1 : Global Vs National Yield of Major Crops

Item	World (kg/ha)	India (kg/ha)	Next to
Paddy	4602	3848	China (6917), Brazil (6210)
Wheat	3531	3219	Germany (7644), France (6757)
Maize	5755	3115	USA (11084), Argentina (7576)
Pulses	1009	664	Russia (2008), Canada (1964)
Sugarcane	70891	69735	Gautemala (121012), USA (82412)
Groundnut	1686	1732	USA (4566), China (3709)
Tobacco	1843	1711	Pakistan (2368)

Source: FAOSTAT

11.3 IMPACT OF COVID-19 ON FARM GATE PRICES: STATE LEVEL

The impact of COVID-19 on farm gate prices at State level in different sub sectors are discussed as under:

11.3.1 Agriculture

The impact of COVID-19 has been fairly uneven on the prices of the agriculture sector at the state level. Some states like Arunachal Pradesh (15%), Mizoram (13.6%), Himachal Pradesh (8%) and Jammu & Kashmir (7%) have reported an increase in the prices of agricultural commodities. On the other hand, states like Karnataka (15%), Telangana (11.7%) and West Bengal (9.7%) have reported a decline in the prices of agricultural commodities.

11.3.2 Horticulture

The impact of COVID-19 is uneven on the prices of horticulture sector commodities. Some states like Arunachal Pradesh (15%), Kerala (13%) and Mizoram (10.7%) have reported an increase in the prices of horticulture commodities. Whereas, states like Karnataka (23%), Tamil Nadu (15.8%), Telangana (15%) and Madhya Pradesh (13.3%) have reported a decline in prices of horticulture commodities. At the aggregate all-India level, there was a 7.6% decline in prices of horticulture products.

11.3.3 Poultry

Poultry prices had reported a significant decline across most of states in the country. Haryana (37.2%), Madhya Pradesh (34.2%), Bihar (31.9%) and Punjab (28.2%) had reported the most significant fall in the prices of the poultry sector.

11.3.4 Dairy

Overall aggregate prices in the dairy sector fell moderately by 5.6%. The states of West Bengal (13.8%), Uttarakhand (15.0%), Jharkhand (14.2%) and Chhattisgarh (11.9%) reported the highest decline, whereas the smaller states of Arunachal Pradesh 25%), Mizoram (7.5%) and Meghalaya (6.7%) showed an increase in the prices of dairy products.

11.3.5 Fisheries

Overall aggregate prices fell moderately in the fisheries sector by 4.8%. COVID-19 had a fairly uneven impact on the fisheries sector prices at the state level. States like Punjab (21.7%), Madhya Pradesh (19.8%), Haryana (19.3%) and Uttar Pradesh (10.7%) reported a significant decline in the prices, whereas, states like Kerala (24.15%), Goa (15%) and Tripura (17.5%) reported a sharp increase in the prices.

11.3.6 Pig/Sheep/Goat

Prices in this sector witnessed only a small overall decline of 2.9% at the all-India level. The states of Haryana (21%), Madhya Pradesh (18.6%), Himachal Pradesh (15%) and Punjab (14.4%) were the ones with sharpest decline in prices, whereas, Nagaland (25%), Kerala (16.1%), Tamil Nadu (10.6%) and Telangana (10%) witnessed the sharpest increase in prices.

11.4 IMPACT OF COVID-19 ON AVAILABILITY OF AGRI-INPUTS

The impact of COVID 19 on the availability of agri inputs during the lockdown period has been discussed in the following paragraphs. The overall availability of agriinputs was reported to have declined in 58% of the sample districts and 38% of the total districts surveyed reported no change in the availability of agri-inputs, whereas only 4% districts reported an increase in the availability of Agri-inputs (Fig 1). The feedback on availability and prices of various agri-inputs viz. seeds, fertilizers, pesticides, rentals agricultural machinery, fodder/cattle feed, etc. were obtained to gain greater insights into the agriculture sector during the lockdown period.

The overall magnitude of change in the availability of the agri-inputs in each category (All-India level) has been depicted in Fig.2. The aggregate availability of agriinputs at all-India level was reported to have declined across all subsectors. The sharpest decline was in the availability of fertilizers (11.2%) followed by fodder/cattle feed (10.8%) and rental agricultural machinery (10.6%). Significant decline was also reported in the availability of pesticides (9.8%) and seeds (9.1%). The reasons for decline in availability of inputs were disruption in supply due to restrictions on movement of vehicles, closure of shops and markets, etc.

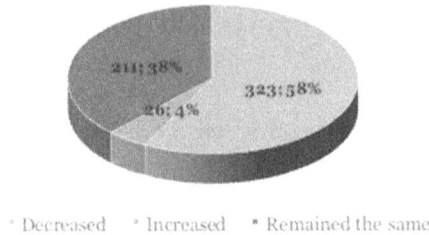

Decreased   Increased   Remained the same

Fig. 1 Number of districts showing change in the availability of Agri-Inputs

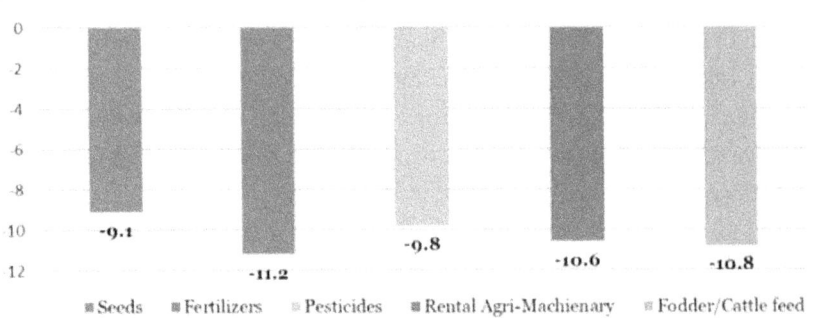

Fig.2: Magnitude of decrease in the availability of Agri-inputs (in%)

11.5 IMPACT ON AVAILABILITY OF AGRI-INPUTS: STATE LEVEL

Although there was a general decline in availability of agri inputs at the national level, yet there were minor variations across states which are discussed as under:

11.5.1 Seeds

The availability of seeds was adversely impacted across all states (except Arunachal Pradesh where seed availability was reported to increase by 2.8%). Nagaland (27.5%), Jharkhand (16.7%), West Bengal (15%), Bihar (14.7%) and Tamil Nadu (12.5%) reported the sharpest decline in the availability of seeds.

11.5.2 Fertilizers

The availability of fertilizers was also significantly impacted due to lockdown imposed owing to the COVID-19 pandemic. The availability of fertilizers decreased in all states except Uttarakhand and Arunachal Pradesh. The states such as Nagaland (35%), Jharkhand (20.8%), Punjab (20%), Andhra Pradesh (18.8%) and West Bengal (18.8%) were all states which reported the largest fall in the availability of fertilizers.

11.5.3 Pesticides

The availability of pesticides also fell sharply across all states in the country except Uttarakhand. The states of Nagaland (35%), Andhra Pradesh (20.6%), Manipur (20%) and West Bengal (18.1%) reported the sharpest fall in the availability of pesticides.

11.5.4 Rental Agricultural Machinery

There was a decline in the availability of Rental Agricultural Machinery across all states in the country due to restrictions on movement of men and material. The states of Nagaland (45%), Jharkhand (18.6%), Assam (17%) and Gujarat (17%) reported the sharpest decline in the availability of Rental Agricultural Machinery.

11.5.5 Fodder/Cattle feed

The availability of fodder/cattle feed also declined across all states in the country due to the COVID-19 pandemic. The states of Manipur (35%), West Bengal (19.7%), Bihar (17.6%) and Jharkhand (16.1%) were some of the states reporting the sharpest decline in the availability of fodder/cattle feed.

The details of the State-wise changes in the availability of agri-inputs across the various subsectors has been provided in Table 2 and 3.

Table 2: State-Wise Impact of COVID-19 on the supply of Agri-inputs

Number of Districts where the supply of Agri-inputs			
State/U.T	Decreased	Increased	Remained the same	Total Districts Covered
Andaman & Nicobar	2	0	1	3
Andhra Pradesh	7	1	5	13
Arunachal Pradesh	9	0	0	9
Assam	15	1	0	16
Bihar	28	2	7	37
Chhattisgarh	10	0	8	18
Dadra Nagar Haveli	0	0	1	1
Daman & Diu	0	0	1	1
Goa	0	0	2	2
Gujarat	10	1	16	27
Haryana	11	0	12	23
Himachal Pradesh	8	0	4	12
Jammu & Kashmir	10	0	3	13
Jharkhand	13	2	5	20
Karnataka	10	1	15	26
Kerala	11	0	1	12
Madhya Pradesh	27	5	21	53

	Number of Districts where the supply of Agri-inputs			
State/U.T	Decreased	Increased	Remained the same	Total Districts Covered
Maharashtra	19	3	12	34
Manipur	5	1	0	6
Meghalaya	10	0	1	11
Mizoram	7	0	0	7
Nagaland	6	0	2	8
Odisha	12	3	7	22
Puducherry	0	0	1	1
Punjab	6	0	16	22
Rajasthan	9	0	12	21
Sikkim	3	0	0	3
Tamil Nādu	11	0	20	31
Telangana	1	1	6	8
Tripura	2	0	4	6
Uttar Pradesh	40	5	20	65
Uttarakhand	7	0	5	12
West Bengal	14	0	3	17
All India	323	26	211	560

Table 3: State-WiseImpactofCOVID-19 on the supply ofAgri-inputs

State-wiseIncrease/Decrease in magnitude of quantity supplied of Agri-inputs					
State/U.T.	Seeds	Fertilizers	Pesticides	Rental Agri-Machinery	Fodder/ Cattlefeed
Andaman & Nicobar	0	0	0	-55	-50
Andhra Pradesh	-13	-18.8	-20.6	-16.1	-7.5
Arunachal Pradesh	2.8	5	0	5	-3.9
Assam	-5	-5	-11.3	-17	-9.4
Bihar	-14.7	-12.9	-12.9	-16.7	-17.6
Chhattisgarh	-5.7	-8.3	-6.7	-15	-12.1
Dadra NagarHaveli	0	0	0	0	-5
Daman &Diu	0	0	0	0	0
Goa	0	-5	0	0	0
Gujarat	-8.6	-10.8	-7.5	-17	-15
Haryana	-7	-7.5	-7.5	-6.6	-9.7
Himachal Pradesh	-5	-16.1	-9.3	-7	-12
Jammu& Kashmir	-11	-12	-9.4	-1.7	-2.8
Jharkhand	-16.7	-20.8	-15.8	-18.6	-16.1

State-wiseIncrease/Decrease in magnitude of quantity supplied of Agri-inputs					
State/U.T.	Seeds	Fertilizers	Pesticides	Rental Agri-Machinery	Fodder/Cattlefeed
Karnataka	-10.7	-11.9	-12.1	-7.9	-7.8
Kerala	-12	-17	-11	-14.1	-8.3
Madhya Pradesh	-4.6	-10.2	-8.4	-12	-14.2
Maharashtra	-4.6	-5.5	-1.2	-10.2	-5.4
Manipur	-21.7	-25	-20	-8.3	-35
Meghalaya	-9.4	-12.8	-9	-20	-7
Mizoram	-13.6	-20.7	-17.9	-3.6	-13
Nagaland	-27.5	-35	-35	-45	-25
Odisha	-13.1	-10.6	-10	-3.8	-6.1
Puducherry	0	0	0	0	0
Punjab	-3.6	-20	-5	-0.4	-6.4
Rajasthan	-8.8	-11.7	-9.3	-11.2	-2.5
Sikkim	-5	-8.3	-11.7	-5	-8.3
Tamil Nādu	-12.5	-13.6	-9.6	-10.6	-9.2
Telangana	-12.5	-7.5	-10	15	-12.5
Tripura	-5	-5	-5	0	-20
UttarPradesh	-6.3	-6.6	-6.9	-11.6	-5.6
Uttarakhand	-0.6	5	3.6	-5	-16
West Bengal	-15	-18.8	-18.1	-10	-19.7
All India	-9.1	-11.2	-9.8	-10.6	-10.8

11.6 IMPACT OF COVID-19 ON PRICES OF AGRI-INPUTS: STATE LEVEL

Although there was a general increase in prices of agri-inputs at the national level, yet there were minor variations across states which are highlighted as under:

11.6.1 Seeds

The prices of seeds had increased across all states in the country. The states of Kerala (15%), West Bengal (13.3%), Tamil Nadu (12%) and Bihar (12%) reported the highest increase in the prices of seeds.

11.6.2 Fertilizers

The availability of fertilizers was also significantly impacted due to the COVID-19 pandemic leading to an increase in prices of fertilizers across all states. The states of West Bengal (16%), Rajasthan (15%) and Bihar (12.4%) reported the sharpest increase in prices of fertilizers.

11.6.3 Pesticides

The price of pesticides also increased across all states in the country due to the shortage in availability. The states of West Bengal (16.1%), Rajasthan (15.8%), and Maharashtra (11.7%) reported the sharpest increase in prices amongst the larger states.

11.6.4 Rental on Agricultural Machinery

The shortage in availability of agricultural machinery due to reduced availability of manpower handling such machines owing to the lockdown also led to an increase in the rent on agricultural machinery across all states. The states of Rajasthan (19.1%), Gujarat (15%), Maharashtra (14.2%) and Bihar (13.2%) reported the steepest increase in rent on agricultural machinery.

11.6.5 Fodder/Cattle feed

The availability of fodder/cattle feed saw the sharpest decline due to the pandemic and thus the sharpest increase in prices was also for Fodder/ Cattle feed. The states of Telangana (25%), Kerala (18.3%), Rajasthan (17.2%) and Himachal Pradesh (17%) reported the sharpest increase in prices of Fodder/Cattle Feed.

The details of the State-wise changes in the magnitude of prices of agri-inputs across the various sub-sectors has been provided in Table 4 and 5.

Table 4: State-wiseImpactofCOVID-19onpricesofagri-Inputs

State/U.T.	Number Of Districts where the prices of Agri-inputs(No.)			
	Decreased	Increased	Remained the same	Total Districts Covered
Andaman & Nicobar	0	2	1	3
Andhra Pradesh	0	6	7	13
Arunachal Pradesh	0	0	9	9
Assam	1	11	4	16
Bihar	0	31	6	37
Chattisgarh	2	10	6	18
Dadra NagarHaveli	0	0	1	1
Daman &Diu	0	0	1	1
Goa	0	0	2	2
Gujarat	3	6	18	27
Haryana	1	14	8	23
Himachal Pradesh	0	7	5	12
Jammu& Kashmir	0	8	5	13
Jharkhand	2	14	4	20
Karnataka	2	10	14	26
Kerala	0	2	10	12

Number Of Districts where the prices of Agri-inputs(No.)			
State/U.T.	Decreased	Increased	Remained the same	Total Districts Covered
MadhyaPradesh	2	26	25	53
Maharashtra	3	22	9	34
Manipur	0	6	0	6
Meghalaya	0	7	4	11
Mizoram	0	7	0	7
Nagaland	0	0	8	8
Odisha	0	13	9	22
Puducherry	0	0	1	1
Punjab	0	8	14	22
Rajasthan	1	15	5	21
Sikkim	0	3	0	3
Tamil Nādu	1	16	14	31
Telangana	0	3	5	8
Tripura	0	1	5	6
UttarPradesh	4	35	26	65
Uttarakhand	1	5	6	12
West Bengal	1	12	4	17
All India	24	300	236	560

Table 5: State-wiseImpactofCOVID-19 on prices of agri-Inputs

State-wise Increase/Decrease in magnitude of Prices of Agri-inputs(%)				
State/U.T.	Seeds	Fertilizers	Pesticides	Rental Agri-Machinery	Fodder/Cattle feed
Andaman&Nicobar	0	0	0	15	20
AndhraPradesh	11	5	5	5	6.8
ArunachalPradesh	0	0	0	0	15
Assam	6.5	8.8	5.8	5	11.2
Bihar	12	12.4	9.8	13.2	12.9
Chattisgarh	4.3	3	3.5	6.8	7.3
DadraNagarHaveli	0	0	0	0	5
Daman&Diu	0	0	0	0	0
Goa	0	0	0	0	0
Gujarat	2.3	5	5	15	11.3
Haryana	6.5	6.1	6.8	8.2	8.9
HimachalPradesh	10.7	10.7	10	10.7	17
Jammu&Kashmir	13.8	10	10.7	8.8	8.8

State-wise Increase/Decrease in magnitude of Prices of Agri-inputs(%)					
State/U.T.	Seeds	Fertilizers	Pesticides	Rental Agri-Machinery	Fodder/Cattle feed
Jharkhand	9	11.3	9.3	8.8	16
Karnataka	6.4	8.6	7.1	3.5	5
Kerala	15	10	10	13.3	18.3
MadhyaPradesh	8.3	9.5	9.1	14.1	9.6
Maharashtra	10.4	10.6	11.7	14.2	11.9
Manipur	11.7	11.7	11.7	10	18.3
Meghalaya	13.6	13.6	12.1	5	20
Mizoram	12.1	16.4	12.1	7.9	11
Nagaland	0	0	0	0	0
Odisha	8.1	5.7	2.3	-2.7	7.7
Puducherry	0	0	0	0	0
Punjab	5	7	9	8.8	7.2
Rajasthan	11.4	15	15.8	19.1	17.2
Sikkim	1.7	0	0	0	5
Tamil Nādu	13	10.6	10	10	12
Telangana	0	0	0	11.7	25
Tripura	5	0	0	0	15
UttarPradesh	6.2	9.4	7.9	10.4	10.2
Uttarakhand	7.5	10	5	8.3	12.8
WestBengal	13.3	16	16.1	13.8	15.6
AllIndia	8.8	10	9	10.4	11.6

11.7 IMPACT ON MSMES

⊙ Micro, Small and Medium Enterprises (MSMEs) are considered to be the backbone of the Indian economy. It is the second largest employment generating sector (after agriculture), employing nearly 120 million people. It contributes over 40% of the overall exports from India. An analysis of the impact of COVID-19 on rural economy is incomplete without considering its impact on MSMEs.

⊙ On consolidating the responses received, we find that the aggregate impact on the economy has been 'Medium. In terms of the state-level analysis, we find that some of the states such as Arunachal Pradesh and Sikkim reported the impact to be 'Low', while other states/UTs such as Haryana, Jammu & Kashmir, Kerala and West Bengal reported the impact to have been 'High'. The state-level results have been given in Table 13.3 in the Annexure.

⊙ The summary of reported responses regarding impact of COVID-19 on MSME sector in sample districts (Fig 3) are given below:

- **Price of Key Raw Materials:** Price of key raw materials was reported to increase or adversely impacted in nearly 46% of the sample districts. Increase in prices of raw materials may have been observed due to restriction of movement of goods during the lockdown, thereby reducing their supply. State-wise analysis of the data shows that the major states which reported a higher proportion of districts with increased prices of raw materials include Andhra Pradesh (69%), Bihar (68%), Rajasthan (67%), Jharkhand (65%) and Haryana (61%).

- **Production Level:** Production levels were reported to have decreased or adversely affected in nearly 97% of the sample districts. Since an overwhelming majority of the enterprises were reported to face challenges in the form of restricted movement of goods and people, reduced access to credit, lower sales, etc., their production may have reduced. State-wise analysis of the data showed that most of the bigger states, including Andhra Pradesh, Chhattisgarh, Haryana, Madhya Pradesh and Tamil Nadu, reported a decrease in production level in all of their sample districts.

- **Cash Flow:** Cash flow constraints were reported in nearly 80% of the sample districts. Reduced purchasing power of other firms/individuals owing to restrictions imposed during lockdown had impacted the cash flow of the MSMEs adversely. State wise analysis of the data showed that the bigger states which reported a higher proportion of districts with increased prices of raw materials include Kerala (100%), Maharashtra (91%), Punjab (91%), Gujarat (89%) and Haryana (87%).

- **Employment:** Among various operational aspects of MSME, employment was reported to be most adversely affected in nearly 96% of the sample districts. The reduced sales, uncertainty about future business prospects and declining financial viability of the enterprises may have forced the enterprises to reduce employment. The State-wise analysis of the data showed that most of the bigger states, including Andhra Pradesh, Chhattisgarh, Haryana, Kerala, Punjab, Rajasthan and Uttarakhand, reported a decrease in production level in all of their sample districts.

- **Supply Chain:** Supply Chain disruptions were reported in nearly 80% of the of the sample districts mainly due to disruption in entire chain owing to the lockdown restrictions. State-wise analysis of the data shows that the bigger states which reported a higher proportion of districts with increased prices of raw materials include Kerala (100%), Haryana (96%), West Bengal (94%), Punjab (91%) and Jharkhand (90%).

- **Export:** Exports, wherever applicable, were reported to have been adversely affected in nearly 86% of the districts. Reduced means of international transportation, reduced foreign incomes and a push by many economies to encourage domestic production may have adversely impacted the exports. State-wise analysis of the data showed that the bigger states which reported a higher proportion of districts with increased prices of raw materials include Kerala (100%), Karnataka (96%), Punjab (96%), Tamil Nadu (97%), Maharashtra (94%), Chhattisgarh (94%), Uttar Pradesh (94%) and Haryana (91%).

- **Consumer Sentiment/Demand:** Consumer sentiment/demand was reported to be adversely affected in nearly 85% of the sample districts. Reduced employment, lower earnings of individuals/households and growing uncertainty may have prompted many households to postpone non-essential expenditure, leading to decline in demand. State-wise analysis of the data shows that the bigger states which reported a higher proportion of districts with increased prices of raw materials include Kerala (100%), Punjab (100%), Haryana (96%), Jharkhand (95%), West Bengal (94%), Madhya Pradesh (94%) and Bihar (91%).

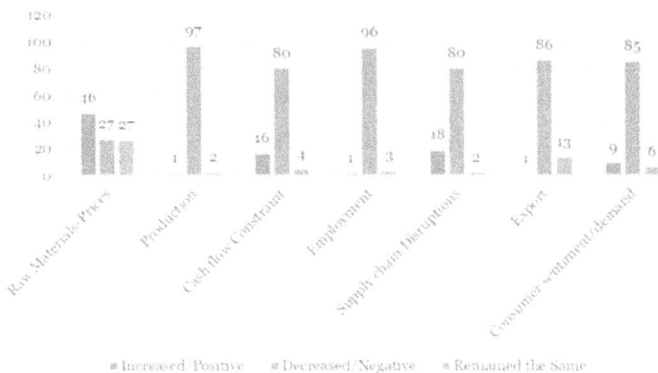

Fig.3 Impact of Key Indicators of MSMEs (Proportion of Districts)

11.8 AT A GLANCE

- ◉ On the whole, at the national level the impact of COVID-19 and the resultant lockdown had been quite harsh on agriculture and allied sector in majority of districts. Among various subsectors, rabi crops were least affected as its harvesting was on the verge of completion but allied sectors such as poultry, fisheries and pig/goat/sheep sector witnessed a drastic fall in demand due to misplaced rumours leading to declining production as well as declining farm gate prices. However, prices of agriculture inputs were estimated to be rising mainly due to disruption in supply chain and closure of shops and markets. Although banking activities were exempted from lockdown, yet basic banking services viz, loans, deposit and recovery were severely hampered in majority of the sample districts in the country. However, the silver lining was the increase in digital banking transactions in majority of the sample districts.

◉ The microfinance sector and MSME sector were the biggest casualty with disruption in more than four-fifths of the sample districts thereby seriously hampering the livelihood in the unorganised sector which provides maximum employment in the rural areas. The activities of FPOs and FCs also came to complete halt. However, these rural institutions including SHGs grabbed the opportunities provided by the situation of stitching face masks, PPEs and preparation of sanitizers thereby helping the society as also earning some income for their members. Further, FPOs in close coordination with local administration in some of the districts were quite instrumental in door-to-door delivery of fruits, vegetable and dry rations to the needy there by extending a helping hand to the society. These rural institutions like SHGs and FCs were also active in creation of awareness in rural areas about COVID 19 and its preventive measures.

BIBLIOGRAPHY

⊙ Ackoff, R.L. (1973). 'Science in the Systems Age: Beyond IE, OR, and MS', Operations Research 21(3): 661-671.

⊙ Ackoff, R.L. and F.E. Emery (1972). On Purposeful Systems, Tavistock, London.

⊙ Agricultural Economic Analysis: John Wiley and Sons, Inc., London

⊙ Ashby, J.A. and L. Sperling (1995). 'Institutionalizing Participatory, Client-driven Research and Technology Development in Agriculture', Development and Change 26(4): 753-770.

⊙ Axinn, N.H. and G.H. Axinn (1983). Small Farms in Nepal: A Farming Systems Approach to Description, Rural Life Associates, Kathmandu.

⊙ Boulding, K.E. (1956). 'General Systems Theory - The Skeleton of Science', Management Science 2: 197-208.

⊙ C.E.BISHOP, W.D TOUSSAINT,., NEWYORK,1958,   Introduction to

⊙ Cederroth, S. (1995). Survival and Profit in Rural Java: The Case of an East Javanese Village, Curzon Press, Richmond.

⊙ Chambers, R. (1983). Rural Development: Putting the Last First, Longman, London.

⊙ Chambers, R. and B.P. Ghildyal (1985). 'Agricultural Research for Resource-poor Farmers: The Farmer-first-and-last Model', Agricultural Administration 20(1): 1-30.

⊙ Checkland, P.B. (1981). Systems Thinking: Systems Practice, Wiley, Chichester.

⊙ Clayton, E. (1983). Agriculture, Poverty and Freedom in Developing Countries, Macmillan Press, London.

⊙ Dillon, J.L. (1980). The Definition of Farm Management', Journal of Agricultural Economics 31(2): 257-258.

◉ Dillon, J.L. (1992). The Farm as a Purposeful System, Miscellaneous Publication No. 10, Department of Agricultural Economics and Business Management, University of New England, Armidale.

◉ Dillon, J.L. and J.B. Hardaker (1993). Farm Management Research for Small Farmer Development, FAO Farm Systems Management Series No. 6, Food and Agriculture Organization of the United Nations, Rome.

◉ Dillon, J.L. and J.R. Anderson (1990). The Analysis of Response in Crop and Livestock Production, 3rd edn, Pergamon Press, Oxford.

◉ Dixon, J.M., M. Hall, J.B. Hardaker and V.S. Vyas (1994). Farm and Community Information Use for Agricultural Programmes and Policies, FAO Farm Systems Management Series No. 8, Food and Agriculture Organization of the United Nations, Rome.

◉ Duckham, A.N. and G.B. Masefield (1970). Farming Systems of the World, Chatto and Windus, London.

◉ FAO (1989). Farming Systems Development: Concept, Methods, Applications, Food and Agriculture Organization of the United Nations, Rome.

◉ FAO (1990). Guidelines for the Conduct of a Training Course in Farming Systems Development, FAO Farm Systems Management Series No. 1, Food and Agriculture Organization of the United Nations, Rome.

◉ Fresco, L.O. and E. Wesphal (1988). 'A Hierarchical Classification of Farm Systems', Experimental Agriculture 24: 399-419.

◉ Friedrich, K.-H. (ed.) (1992). Readings in Farming Systems Development, Food and Agriculture Organization of the United Nations, Rome.

◉ Grigg, D.B. (1974). The Agricultural Systems of the World: An Evolutionary Approach, Cambridge University Press.

◉ Heady Earl O and Herald R. Jenson,1954, Farm Management Economics:, Prentice Hall, New Delhi,

◉ Heady, Earl O, 1964, Economics of Agricultural Production and Resource Use:, Prentice Hall of India, Private Limited, New Delhi

◉ I.J. Singh,1976, Elements of Farm Management Economics: Affiliated East-

◉ Kast, F.E. and J.E. Rosenzweig (1974). Organization and Management: A Systems Approach, 2nd edn, McGraw-Hill Kogakusha, Tokyo.

◉ Kostrowicki, J. (1974). The Typology of World Agriculture. Principles, Methods and Model Types, International Geographic Union, Warsaw.

◉ Makeham, J.P. and L.R. Malcolm (1986). The Economics of Tropical Farm Management, Cambridge University Press.

- Matlon, P., R. Cantrell, D. King and M. Benoit-Cattin (eds) (1984). Coming Full Circle: Farmers' Participation in the Development of Technology, IDRC, Ottawa.

- McConnell, D.J. (1972). The Structure of Small Farms in Peshawar, NWFP, Pakistan, UNDP-FAO Consultant's Report TA3070, Food and Agriculture Organization of the United Nations, Rome.

- McConnell, D.J. (1992). The Forest-garden Farms of Kandy, Sri Lanka, FAO Farm Systems Management Series No. 3, Food and Agriculture Organization of the United Nations, Rome.

- Mikkelsen, B. (1995). Methods for Development Work and Research: A Guide for Practitioners, Sage Publications, London.

- Norman, D.W. (1980). The Farming Systems Approach: Relevancy for the Small Farmer, Rural Development Paper No. 5, Michigan State University, East Lansing.

- Norman, D.W., F.D. Worman, J.D. Siebert and E. Modiakgotia (1995). The Farming Systems Approach to Development and Appropriate Technology Generation, FAO Farm Systems Management Series No. 10, Food and Agriculture Organization of the United Nations, Rome.

- Rhoades, R.E. and R.H. Booth (1982). 'Farmer-back-to-farmer: A Model for Generating Acceptable Agricultural Technology', Agricultural Administration 11(1): 127-137.

- Rhoades, R.E. and R.M. Booth (1982). 'Farmer-back-to-farmer: A Model for Generating Acceptable Agricultural Technology', Agricultural Administration 11(2): 127-137.

- Ruthenberg, H. (1976). 'Farm Systems and Farming Systems', Zietschrift für Ausländische Landwirtschaft 15(1): 42-55.

- Ruthenberg, H. (1980). Farming Systems in the Tropics, 3rd edn, Oxford University Press.

- S.S. Johl, J.R. Kapur, 2006, Fundamentals of Farm Business Management:, Kalyani Publishers, New Delhi

- Sankhayan, P.L.,1983, Introduction to Farm Management: Tata – Mc Graw – Hill Publishing Company Limited, New Delhi,

- Shaner, W.W., P.F. Philipp and W.R. Schmehl (1982). Farming Systems Research and Development: Guidelines for Developing Countries, Westview, Boulder.

- Spedding, C.R.W. (1979). An Introduction to Agricultural Systems, Applied Science Publishers, London.

- Subba Reddy, S., Raghu ram, P., Neelakanta Sastry T.V., Bhavani Devi I.,2010, Agricultural Economics, Oxford & IBH Publishing Co. Private Limited, New Delhi

⦿ Tripp, R. (ed.) (1991). Planned Change in Farming Systems, Wiley, Chichester.

⦿ Upton, M. (1973). Farm Management in Africa, Oxford University Press, London.

⦿ Upton, M. and J.M. Dixon (eds) (1994). Methods of Micro-level Analysis for Agricultural Programmes and Policies, FAO Farm Systems Management Series No. 9, Food and Agriculture Organization of the United Nations, Rome.

⦿ Von Bertalanffy, L. (1973). General System Theory: Foundations, Development, Applications, Penguin, Harmondsworth.

⦿ Walker, T.S. and J.G. Ryan (1990). Village and Household Economies in India's Semi-arid Tropics, Johns Hopkins University Press, Baltimore.

⦿ Werner, J. (1993). Participatory Development of Agricultural Innovations: Procedures and Methods of On-farm Research, Schriftenreihe der GTZ No. 234, GTZ, Eschborn and SDC, Bern.

⦿ West press, Private Limited, New Delhi.

INDEX

A

Aberrant 166

Ability 82, 95, 238

Access 10, 109, 235

Adaptation 88-89, 98-99, 200

Analysis 58-63, 77, 80, 92-93, 138, 147, 189, 255

Animals 30-31, 65-67, 168-169, 185, 211

Apiculture 32

Approaches 88-89

Artificial 32, 38

Asymmetry 112

B

Bamboo 56

Barley 57

Beehives 32

Beeswax 32

Beetles 196

Biodynamic 225

Breeding 31, 168-169

Breeds 31

Budgets 148

Buildings 66-67

C

Capital 9, 44, 105, 112-120, 125-126, 129-130, 157

Carbon 191, 210

Cattle 249, 252

Chemicals 230

Climate 52, 188-189, 191-193, 195-200, 203-205, 215

Coffee 110

Collection 152, 175, 240

Common 32, 119

Community 46, 108, 122, 219, 222

Component 42

Compost 97, 228

Consumer 149-151, 154, 162

Consumers 107

Covid- 246

Cropping 98-101

Customer 72, 138, 143, 145-146, 148, 162

D

Decline 25, 246-249

Depression 31

Diagnosis 63

Direction 76

Diseases 224

Dominance 140

Drainage 9

Drought 84

E

Ecological 242

Economics 58, 126, 135, 259

Ecosystems 190-191

Education 18, 112, 119

Elements 38-39, 46-47, 51

Energy 60, 177, 179, 222

Erosion 10

F

Family 59, 65, 74

Famine 236

Farmers 7, 13-16, 18, 23, 40, 46, 56, 70, 85, 87-88, 99, 103-104, 106-108, 111, 113, 152-153, 156, 158, 191, 214, 216-217, 219-222, 224-226, 229-231, 233, 238

Farmland 4-5, 8, 11-12

Feeding 240-241

Fertility 7-8, 55

Fisheries 33, 121-122, 247

Fodder 249, 252

Forage 187

Forestry 121

Formalisms 89

G

Global 37, 189, 198, 200, 205, 217, 246

Grains 185, 220

Grazing 82-83

Greenhouse 188

Growers 229

Growth 76, 79, 111, 114-117, 141, 152, 158, 184, 237, 245

H

Harvest 187

Health 156, 229

Honeybees 32

I

Inbreeding 30-31

Infants 240-241

Innovation 71, 140

Insects 196, 210

I

Irrigation 10, 53, 62, 100-101, 184, 187, 202, 205

L

Livestock 36-37, 44, 58, 65, 165, 168, 211, 227-229

Lockdown 255

M

Machinery 249, 252

Manures 184

Marginal 25-26, 133-135

Market 6, 13, 70, 77, 89, 95, 108, 110, 134, 138-140, 144, 147, 149-153, 160-162, 231

Marketers 138

Millet 57

Mustard 57

N

Nitrate 215

Nitrogen 210

Nutrients 184, 210, 232

Nutrition 240

O

Operations 18, 170, 173

Organic 208-214, 216-234

P

Pathogens 197

Pesticides 186, 209, 217, 220, 224, 230, 232, 249

Phenology 197

Pollinator 225

Poultry 30, 66, 247

Prototype 180

R

Rainfall 84, 99, 201

Rajasthan 252

Research 21, 102-103, 113, 116, 139, 258

Residual 84

Resistance 186

Revenue 29

Revolution 220

Rotations 228

S

Seedbed 185

Soybeans 196

Spatial 88

Species 31

Standards 217, 219

Storage 67, 163, 187

Summit 239

Survey 109

T

Textile 150

Tropics 200-204

V

Valuable 226

W

Waterways 215

Weather 166